Oil and Natural Gas Exploration and Drilling Operations

FUNDAMENTALS OF INVESTING IN OIL AND GAS SERIES

Chris Termeer

First Edition
Manufactured in the United States of America

Edited by Mike Rutz

For Information pertaining to the use of copyrighted material, please contact the following:

Chris Termeer
PO Box 3443 Clearwater Beach Florida 33767
www.christermeer.com/chris-termeer-publishing
or fill out contact form on Christermeer.com
727-475-1300

Termeer, Chris.
Fundamentals of Investing in Oil and Gas
Includes index.
ISBN: 978-0-9890434-4-1 (pbk.)

Library of Congress Control Number: 2013946420

Preface

Oil and Natural Gas Exploration and Drilling Operations is from the series of "Fundamentals of Investing in Oil and Gas" and will be a light-to-intermediate read intended for those who already have a preexisting understanding of oil and gas history, common oil and gas terms, legal documentation, markets, land valuation, legal documentations, government and state requirements, market trends and investment risks. If you are not familiar with these topics then this book may not be as useful as the first book I published called "Fundamentals of Investing in Oil and Gas" which is a large red book 8.5 x 11".

This book is an isolation of content specifically about exploration and drilling operations from the book "Fundamentals of Investing in Oil and Gas" and, though the content is nearly identical, it is intended to be a handy and quick reference for individuals that are actively participating in exploration or drilling activities.

Table of Contents

Chapter 1 – How Oil Occurs in Nature

"Formula for success: rise early, work hard, strike oil."
- J. Paul Getty

Oil and natural gas are both considered fossil fuels as they are formed by fossilized remains of plants and animals. It is widely believed that earth's earliest lifeforms lived and died in ancient oceans and seas. As time passed, their remains accumulated on the seafloor as organic sediment. Gravity compressed this organic sediment into thick, heavy and dense geological layers. This process continued until a dense hydrocarbon substance known as kerogen was formed. With the passage of enough time, kerogen was compressed into what is called source rock, the basic fundamental requirement of an oil reservoir.

	Source	Creates
Type I	Ancient sea-based life forms (algae, etc)	Predominantely liquid hydrocarbons
Type II	Ancient sea and land-based life forms	Liquid and gasous hydrocarbons
Type III	Ancient land-based life forms (plants)	Predominantely solid hydrocarbons (coal)

Table 1 – 1
Kerogen Types

Geological forces, such as shifting tectonic plates and volcanic activity, reshaped these source rock layers, causing them to move up, down, sideways or even fold up on top of each other. Earth's constantly changing geology not only converted kerogen into oil over millions of years but also trapped some of it within a porous rocky material capped beneath an additional layer of impenetrable rock.

Today's industry uses a series of terms that are often misinterpreted. A formation is simply any identifiable layer of rock named after its geographical location and dominant rock type that may or may not contain hydrocarbons. A reservoir is a rock formation that contains oil and/or natural gas trapped beneath a cap rock layer. A formation which includes an oil reservoir is then called an oil formation.

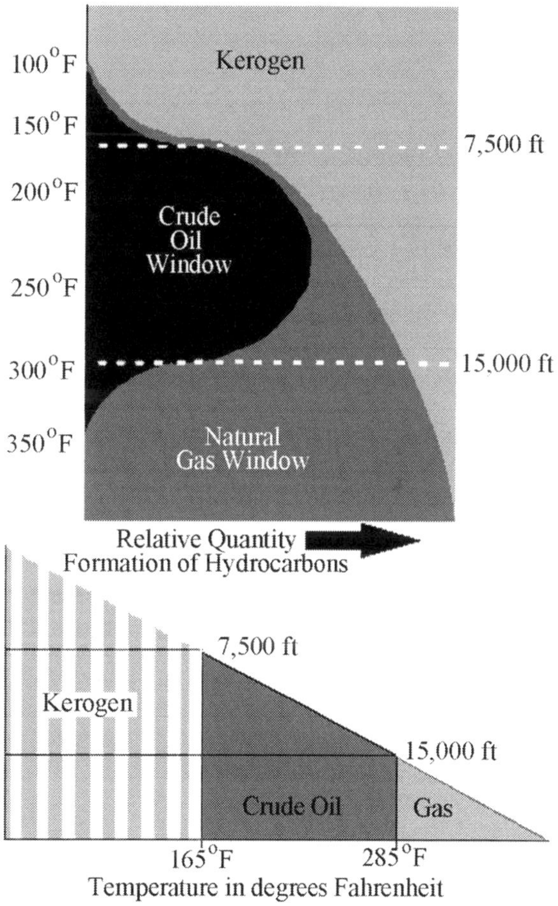

Figure 1 – 1
Hydrocarbon Formation Windows

Consider the probability of flipping a coin and landing on heads five times in a row. Similarly, five rare events must occur to form oil, each step making the cumulative process less likely to occur. It is this combination that makes this substance so rare and valuable.

1) Organic material accumulates in dark oxygen poor areas, known as dead zones, at the bottom of a lake, delta, sea, or ocean where it is left undisturbed for long periods of time to create kerogen.
2) Kerogen must be compressed into a hydrocarbon rich rock-like structure known as a "source rock."
3) Geological forces must move the source rock into the "oil window" located deep underground. The oil window is a precise range of conditions (depth and temperature) necessary for oil to form.
4) The source rock must be buried in this oil window for a sufficient length of time.
5) A layer of impermeable rock must cap the source rock beneath the surface to prevent any substances from escaping. Cap rocks also trap additional pressure which can lead to increases of heat within the formation as oil is formed.

One of the reasons why oil reservoirs exist below the depths at which they are created (15,000+ feet) is the continuation of the same geological forces that initially moved them into the oil window. Conversely, shallow reservoirs (less than 7,500 feet) exist because liquids and gases will move from areas of high pressure to low pressure whenever a differential exists.

Reservoir Composition
For the purposes of this discussion, we will be focusing on structures which tend to contain crude oil. One of the more common oil formation types is sandstone, a rocky material comprised mostly of highly compressed sand. The oil within these structures is not a continuous lake or pool, as commonly perceived, but is instead located within the spaces between the individual rocks, known as pores.

Figure 1 – 2
Microscopic View of Pores
Source: Department of Energy (DOE)

Porosity
Most formations or rock structures contain small voids of space between the particulates of sand or rock. Porosity, one of the most important measurements when studying a formation for oil or natural gas, measures the percentage of this void space. It can be thought of as the volume of empty space divided by total volume of the formation. A good oil formation will have above 18% porosity to be economical without needing further treatment or stimulation.

Permeability

Otherwise referred to as K, permeability measures the degree to which a liquid or gaseous substance can flow through a solid. This happens when multiple porous layers are piled on top of each other. The pores of each layer will sometimes align in a way that creates channels throughout the formation through which liquids and gases can flow.

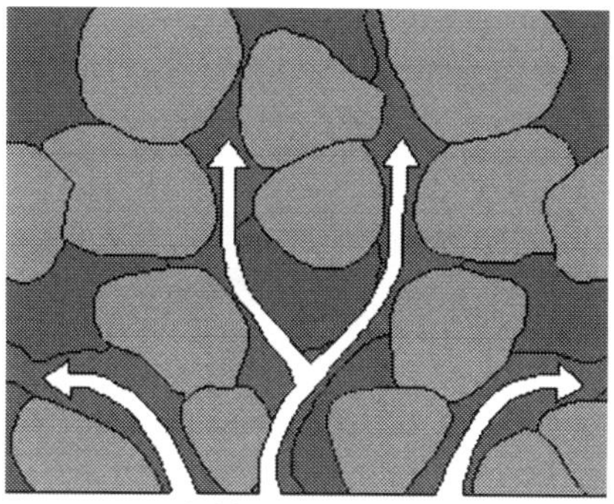

Figure 1 – 3
Example of Formation Permeability

In the oil and gas industry, a unit called a millidarcy (mD) is used to quantify a formation's permeability. The higher the number of millidarcies, the more feasible it should be to extract oil. A K rating above 10mD is considered a good prospect for producing oil.

Types of Oil Reservoirs

Oil and gas reservoirs can be classified by their distinctive shapes. Different types of reservoirs have different implications for drilling and production. A trap, a geological formation that prevents lateral and vertical movement of a gas or liquid, is an important indicator of potential oil reserves. Without a trap, hydrocarbons are able to seep up to the surface. The two main trap types are Structural and Stratigraphic. Structural traps are further classified as being Anticlinal, Fault, or Salt Dome. Stratigraphic traps can be Pinchout, Unconformity or Lens and Reef.

Structural Traps

Usually formed by tectonic actions, structural traps are reinforced by the compression forces of gravity. The impermeable rocks above and below the formation are the distinguishing features of a structural trap.

Anticline Traps

Anticlinal traps are created when multiple layers of rock, and their contained liquids, bend into a dome shape. The contents of these structures are arranged by density, with the least dense substance at the top. This type of structural trap contains a great proportion of the world's known oil reserves.

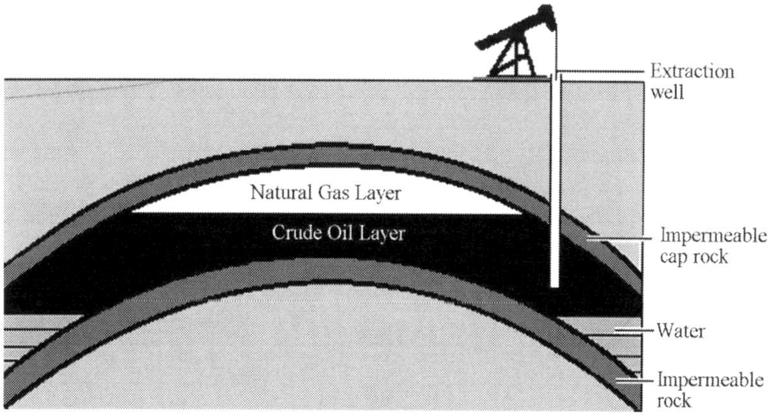

Figure 1 – 4
Anticline Trap

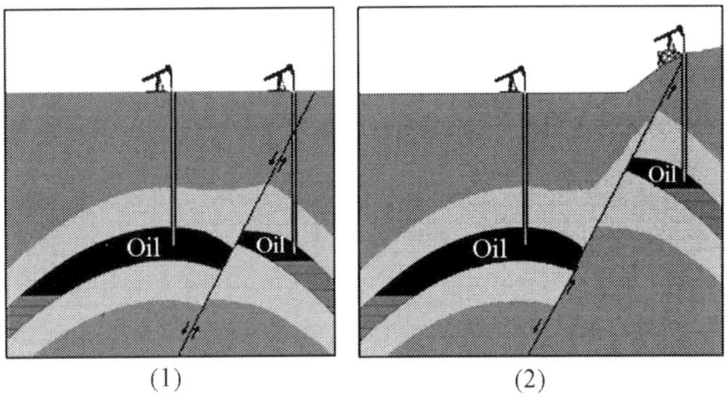

Figure 1 – 5
Left – Fault Trap Early Migration Stage
Right – Fault Trap Mature Migration Stage

Fault Traps

Faults are created when layers of the Earth's outer crust collide or shift into each other. Many of the Earth's surface features are created by fault activity, including mountains and canyons. Fault traps are similar to anticlinal traps with the exception of having a nearby fault line.

On the left of Figure 1 – 5, only slight migration occurred within the reservoir as oil and water can move freely under optimal permeability and porosity conditions. As depicted on the right of Figure 1 – 5, this movement continued, in effect "pinching off" and isolating a small oil reservoir from the main, long before any oil development occurred. Over time, the distance between these two structures – which were once combined – can be quite large.

Salt Dome Traps

The Spindletop mega-well was one of the first salt dome traps discovered. A salt dome is formed from salt deposits trapped within an ancient body of water. The density differentials between the salt deposits and their surrounding rock cause the salt layer to move toward the surface, resulting in a distinct dome-like structure. The upward thrust of the dome was originally formed by a fluid. After many years, these domes solidify. Salt domes often create donut shaped structures with pockets of oil trapped around the salt protrusion, commonly referred to as plugs. Conversely, the salt dome protrusion itself can contain rich hydrocarbons, as was the case with Spindletop.

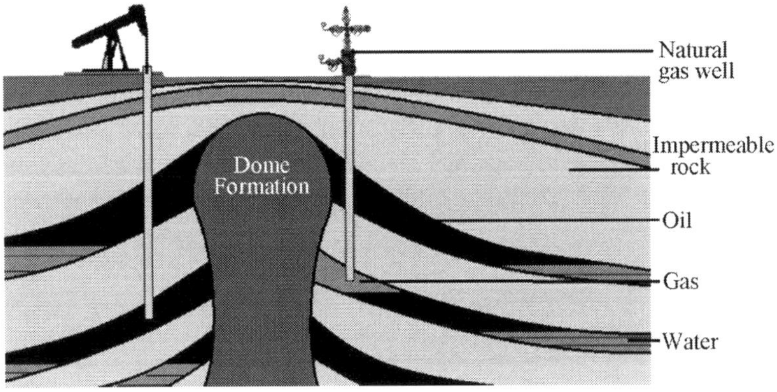

Figure 1 – 6
Salt Dome Trap

Deeper shale or clay layers under similar conditions also form domes, but these protrusions are called shale diapirs. They should not be confused with shale deposits, such as the famous Eagle Ford Shale, located in south Texas, which are composed of a thick layer of rock that can span large distances.

Stratigraphic Traps

The accumulation of sediment layers, called strata, form stratigraphic traps over long periods of time. This is a sharp contrast to the geological processes which create the structural traps discussed previously. Stratigraphic traps include pinchout, uniformity, and lens and reef.

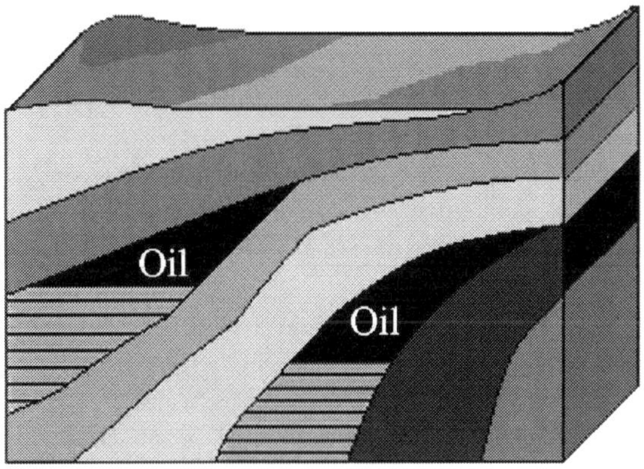

Figure 1 – 7
Pinchout Trap

Pinchout Trap

In a pinchout trap, one or more layers of impermeable rock divide a layer of hydrocarbons. Sometimes, this type of trap is mistakenly interpreted as being two different oil-bearing reservoirs, when, in fact, it is technically only one discontinuous reservoir.

Unconformity Traps

The distinct feature of the unconformity trap is the intersection of perpendicular rock layers, with at least one vertical layer containing a reservoir surrounded by impermeable rock. Millions of years ago, the vertical layers in Figure 1 – 8 used to protrude up through the surface, but were eventually weathered away and later capped by a layer of impenetrable sediment. This intersection creates a "T" shape that gives this geologic structure the nickname "truncated trap."

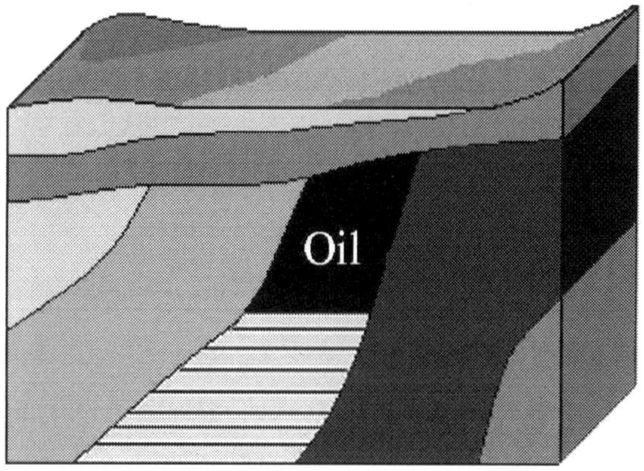

Figure 1 – 8
Unconformity Trap

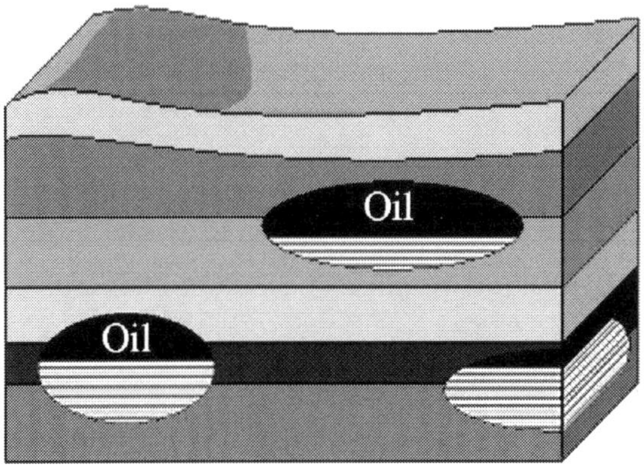

Figure 1 – 9
Lens Trap

Lens and Reef Traps

A lens trap is formed when an impermeable layer of rock combines with a porous and permeable layer of rock containing hydrocarbons. These types of traps are commonly found near the beds of prehistoric rivers, creating pockets of small oil reservoirs. Reef traps, similar to lens traps, are created by small, isolated pockets of porous limestone made up of once-living reefs.

Knowing what types of reservoirs can exist underground is vitally important for the people and companies searching for oil. Different structures call for different techniques during all the phases of oil drilling and production. Identifying which formation types are present underneath a prospect is highly recommended before any development takes place.

Unconventional Sources of Oil and Gas

In addition to the conventional reservoir types previously mentioned, there are a number of so-called unconventional sources of oil and gas that have been getting increased attention, especially in recent years. One of the key indicators of an unconventional source is the lack of differing source, reservoir, and cap rock structures. Unconventional oil sources include oil shale, tight oil and oil sands, while unconventional sources of natural gas are being found in the form of methane hydrates, coalbed methane and shale and tight gas.

Oil shale: formations never entered the hydrocarbon window depicted in Figure 3 – 1. Instead, they contain kerogen, which is much denser and more expensive to refine than conventional crude oil. Kerogen-bearing rock is first strip-mined before the hydrocarbons can be isolated from rock using heat.

Tight oil: is often confused with both oil shale and shale oil by those not familiar with the industry. Tight oil refers to conventional crude oil found in certain types of rock formations once considered unviable for oil production. One interesting feature about these structures is that they are the source rock, the reservoir, and the trap. They are typically found at great depth beneath the surface, are relatively narrow in vertical size and are of extremely low porosity and permeability.

Oil sands: are high density hydrocarbons mixed with sand at relatively shallow depths. As with shale oil, these hydrocarbons never entered the hydrocarbon window. Oil and sand must be separated by one or more energy-intensive processes, either at the surface using heat and centrifuges or at their source below the ground with steam and/or hydrocarbon solvents. Once extracted, the oil is often called natural bitumen, tar sand or bituminous sand because of its chemical and physical similarity to the refined crude oil byproduct called bitumen, often used to make asphalt.

Methane hydrates: refer to methane (CH_4) molecules that are surrounded by, but not chemically bonded to, frozen water. It's been estimated that there are more methane hydrate resources than all other conventional energy sources combined. It is so potentially vast that experts are worried about tapping it out of fear of releasing methane into the atmosphere, a scenario that would exacerbate any known climate model.

Coalbed methane: is methane adsorbed into solid coal. Once considered a safety nuisance during the production of coal, its production occurs in a manner somewhat similar to conventional gas, with a coal seam as the target structure.

Tight gas: is found in rock structures once considered completely impermeable. Fracturing and acidizing are essential to the success of many tight gas wells.

Shale gas: is named after the *type* of rock within which it is found. These structures are notoriously impermeable, but with the continuing improvement of hydraulic fracturing, long-known formations are now economically viable.

As with unconventional oil, unconventional natural gas sources have changed the way industry professionals look at subsurface formations and structures. While higher permeability and porosity are still generally required for traditional vertical oil and gas wells, new options are available if the economics are right. Hydraulic fracturing, horizontal drilling and completions and the additional required upstream processing all increase the cost of developing these unconventional oil and gas sources.

Chapter 2 – Exploration

"Drill for oil? You mean drill into the ground to try and find oil? You're crazy."
- Associates of Col. Edwin Drake refusing his suggestion to drill for oil in 1859

One of the fundamental goals of exploration can be summarized by the old phrase "work smarter, not harder." In the earliest days of the industry, the decision of where to drill for oil was based on finding nearby surface seeps, or places where these valuable hydrocarbons would come to the surface naturally. The majority of these natural seeps were found and extracted during the industry's infancy. Later on, as early pioneers gained more experience, they started to identify surface features as a potential indicator of subsurface reservoirs. The famous Spindletop well was developed solely on the hunch that a hill located in the middle of otherwise-flat terrain meant there was an oil-filled salt dome below, a hunch, which, in this case, proved enormously successful.

A *Wildcatter* is an individual that picks an area of unproven land to drill for oil or gas with little knowledge of what lies below. Typically considered a negative term associated with oil drilling these days, few take into consideration that many of the early pioneers were wildcatters whose efforts contributed greatly to the modern science of geology. Geology is the study of solid earth, the rocks of which it is composed, and the processes through which they were created.

Drilling a hole into the earth has always been a very expensive activity. It requires enormous sums of financial capital, which increases with greater depth. Many wildcat wells are unsuccessful; the ratio of "dry holes" to successful wells tends to be around 6-to-1. As the industry developed, it became necessary to find a more efficient way to locate subsurface formations.

Step-out

To step-out, or stepping out, originated from the practice of drilling an additional well next to a successful one. The step-out technique, relevant to field development, is still used today as a method of developing a reservoir that can support multiple wells and to determine the outer edges of a formation. In a multi-well development project, the first well drilled is called an appraisal or definition well. The purpose of an appraisal well is to determine a formation's potential payzone. A payzone is defined by the following characteristics:

- porosity
- permeability
- presence of hydrocarbon content
- reservoir thickness
- viability of extraction

Once defined, a payzone is typically expressed as the formation's name and associated range of depths. Another form of stepping out would be to re-enter an existing oil field with the intention of drilling deeper in order to find new discoveries. But this form of exploration can be challenging if the existing conditions of the re-entry well are not thoroughly known.

Surrounding Production

The early oil industry was similar to the "gold rush" of the mid-to-late 1800s. Whenever someone "struck gold," others immediately rushed in to snap up surrounding properties with the belief that they would find gold on their land as well. Similarly, when someone hits a good well, others tend to rush in to adjacent land to try and tap the same reservoir, or at least a reservoir of similar potential. A surrounding production analysis is based on this concept, and these publicly available production records are still used today as the foundation for looking at a new prospect.

Often, people that come in and purchase property adjacent to surrounding production lack either additional resources to explore and/or the knowledge needed to successfully extract hydrocarbons, assuming they are even there. These novice speculators often overlook concepts like drainage, which is the movement of fluids from one location to another.

An oil exploration and production (E&P) company has access to public records of when a property was purchased. If a new owner is unwilling to cooperate with a company, by expecting an unreasonable royalty portion for example, the E&P company could move around this landowner and use its financial capital more efficiently. This leads to a "leap frog" approach to development which often results in a drainage-based argument of stolen resources. The law that is most commonly referred to in this scenario relates back to the old English Common Law of Capture.

An example of capture is seen in bird hunting: A bird takes off from the private property belonging to 'owner A' and subsequently lands on another property belonging to 'owner B'. If that bird is then shot by 'owner B', was there a crime committed? Since the bird is not the legal property of 'owner A' and it moved itself from one place to another, it actually becomes property of 'owner B' once it crosses the property line. Relating this back to the oil industry, the disgruntled landowner who may feel that they were a victim of malicious acts of drainage will have an extremely difficult task proving that a crime took place. Other related rights for E&P companies include Riparian water rights, where any landowner whose property is along the path of a body of water has the right to make reasonable use of it.

One of the problems with public records is that they are often misleading as early explorers did not want their activities exposed to the general public. Many records will conveniently lack information needed by other companies looking to move into that area. It is one thing to find oil and natural gas, but it is another to successfully extract it as will be seen in later chapters. Disputes between exploration companies and private landowners occur more often than they should, often impeding further development of valuable resources that can safely and easily be enjoyed by both parties.

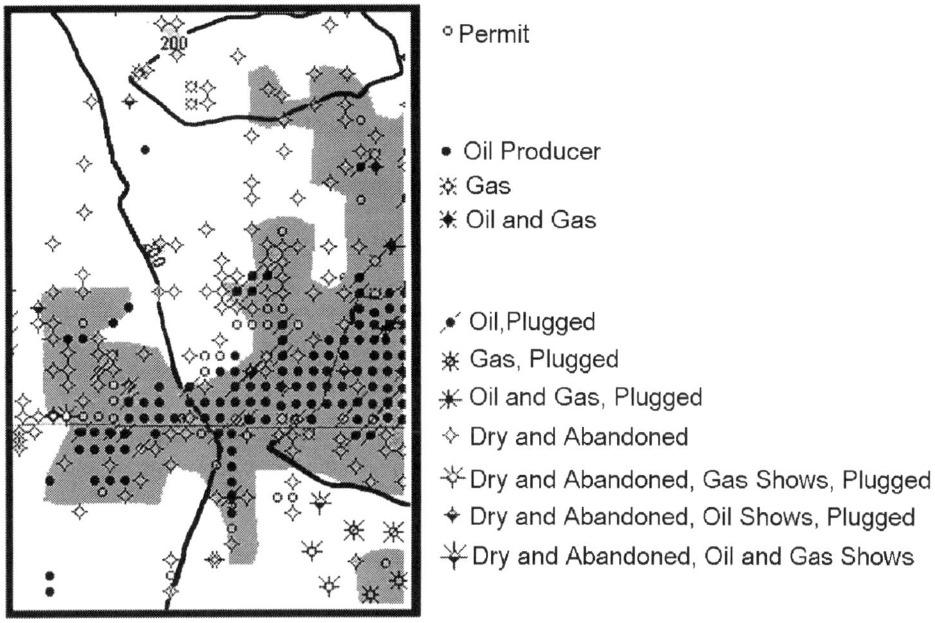

Figure 2 – 1
Example of Surrounding Production Analysis

Drilling Samples
Many people think that exploration activities cease once a new well has broken ground. In fact, these efforts continue throughout the development of a new formation. An expert known as a mudlogger pays close attention to drill cuttings that are brought to the surface. These samples are compiled in a geological column to provide insight as to the characteristics of the target. When developing a new area, the composition and properties of the structures above and below the payzone must also be known as these layers will impact future exploration and production efforts.

If one is planning to develop a new area of land that has not been explored, drilling samples and surrounding production analysis will provide valuable insight to what would otherwise be unknown. The combination of these two methods is extremely powerful and is still used today.

Modern Exploration Techniques
Today, powerful computers and other advanced technologies support the science behind exploration. Over the years, the physical and chemical properties of oil have become better understood. Few realize that crude oil has radioactive and magnetic properties. It is these properties which may be detectable under the correct conditions. There are a variety of ways geologists and geophysicists can evaluate a given area of land to determine the likelihood and magnitude of oil or other hydrocarbons being present. These include magnetic, gravimetric, seismic, and radiometric analyses.

Magnetics

Magnetic surveys gather data from specialized sensors that measure the relative magnetic field strengths caused by minerals in the Earth's upper crust. The Earth's magnetic field is directly influenced by its geological features and structures below. Sensors map sedimentary basin structure, lithology, geological faults, lineaments and fracture systems, along with magnetic anomalies. Magnetic anomalies are usually indicative of possible deposits. These surveys are typically conducted by airplanes or satellites as they can cover large amounts of land in one pass.

One of the problems with magnetic surveys is the fact that some prospective projects lay underneath cattle ranches. Cattle are unique creatures as they tend to eat everything as they graze, including ranching wire and other random sharp metal objects that can destroy their internal organs. To prevent this, ranchers intentionally feed their livestock powerful magnets that help the cattle by attracting and safely storing such hazardous materials within their first stomach. A potential project located on ranching land consisting of large amounts of cattle with magnets in their stomachs can lead to misinterpretation. Many have a hard time believing that cows in this country contain magnets in their stomachs, but all one needs to do is a simple search for cow magnets on the internet to see how large the industry truly is.

One of the other problems with magnetic surveys is that they often cannot determine the depth at which a potential formation may exist. This can lead to a "false positive" reading as multiple non-commercial zones stacked on top of each other can look like a profitable reservoir.

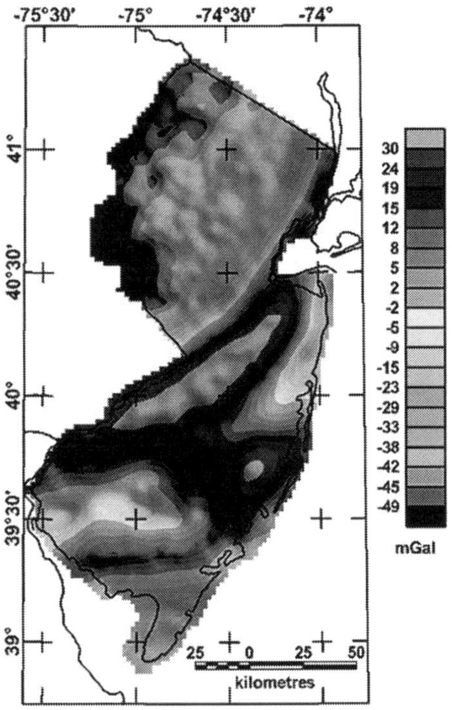

Figure 2 – 2
Sample Gravimetric Survey
Source: U.S. Geological Survey (USGS)

Gravimetric

The measurement of an object's gravitational field is unique throughout nature. Gravimetric surveys analyze the slight changes in density of subsurface rock which has proven useful when determining the presence of a fluid, such as oil.

Like magnetics, these surveys often lack the 3-dimensional ability of showing a target's associated depths. Efforts to "capture" the third dimension involve taking multiple readings at different locations and then trying to cancel out interference with complex filters and computer algorithms. Despite their best efforts, the technology has not developed to the point of being able to make precise formation depth estimates. Instead, they're ideally suited for providing a range of depths that are likely to be worth investigating further, at best.

Radiometrics

Otherwise known as gamma-ray spectrometry, radiometric analysis is an effective geological mapping technique if combined with surrounding production and previous well samples analyses, discussed earlier. If done correctly, radiometrics can be a cost-effective way to determine whether a prospective formation is continuous or isolated.

Radiometrics refers to the measure of decaying atomic particles, particularly gamma radiation, emitted from various types of matter, including the earth's crust. Radiometric technology relies on 3 elements that are present in all gamma radiation: thorium (Th), potassium (K) and uranium (U). One of the known properties of hydrocarbons is that they block uranium's contribution to an area's gamma radiation pattern. Therefore, a radiometric survey conducted for oil exploration looks for a level of uranium which is relatively less than what would be considered normal for the given region.

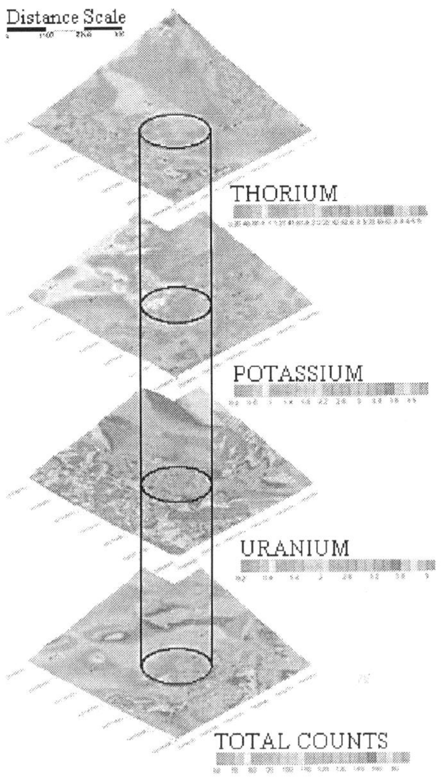

Figure 2 – 3
Sample Radiometric Survey

In terms of radiometric analysis, I find potassium to be a positive sign for finding oil reservoirs. Since we believe hydrocarbons are created from previous plant and animal lifeforms, we can look at how this element is required for metabolic processes for modern plants and animals. Plants today use potassium to build cellulose for cell walls along with the formation of chlorophyll, a precursor for photosynthesis. In addition, potassium is the ninth most common element by mass in the human body. Humans and animals create energy through what is effectively a cellular sodium-potassium pump. Without potassium there would be no plants or animals on our planet.

I would like to think that our prehistoric single-celled ancestors had these metabolic processes that required large amounts of potassium, the same can be said about thorium. Because of this, we see high levels of these elements in crude oil. If potassium and thorium levels are present in a radiometric survey, it simply means that it is a sign or precursor that life could have existed in that time frame. If the structure is correct, there is a chance that hydrocarbons may be found.

Potassium and thorium are also found in many kinds of materials, such as fertilizers and asphalts, whose presence may not have been disclosed or known before the survey. Fertilizer in the United States is typically packaged according to three elements: nitrogen, phosphorous, and potassium. Thorium is found in the phosphorous content of fertilizer and is present in the molecules associated with phosphate. Therefore, if the prospect is located on a field that contains crops, the survey may be incorrect. The ground acts like a sponge in many ways, and it may be difficult to accurately calculate how long fertilizers will be detectable by radiometric equipment. Issues with the ability to calculate accurate depths are also a concern with this type of survey. Radiometric surveys have also attempted to take multiple readings at different angles as a way to display their results in a 3D fashion. But I have a difficult time finding compelling reports that justify drilling a prospect based on these surveys alone.

Seismic

Seismic analysis is one of the intriguing technologies used in the industry today. A seismic report typically comes with a number of bright, multi-colored charts and graphs that try to convey what is going on beneath the ground. As with all the techniques mentioned in this chapter, seismic analysis is a useful geological tool. This information may help facilitate the development of a potential prospect in a way that more efficiently allocates financial resources and minimizes cost for deep formation exploration. It allows for a highly educated guess as to where to drill the first appraisal well within a given field.

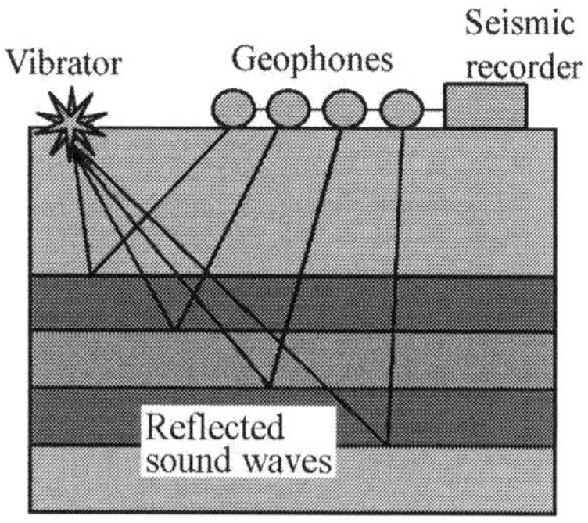

Figure 2 – 4
Basic Seismic Concept

The principles behind seismology are based on the known interactions of sound waves through different layers. Sound waves tend to reflect or refract when encountering materials with different acoustic properties. This tends to happen at points of distinct density changes between formations. Seismic analysis is conducted by sending a signal through the ground and then waiting for it to bounce back to be detected by geophones. Computers then analyze the collected data to create an estimate of how the formation might be shaped.

Refraction is a scientific term for a change in a wave's direction as it passes through materials of differing densities. Sound refraction is similar to light refraction. Placing a pencil in a glass of water as seen in Figure 2 – 5 best demonstrates this concept. The pencil appears to be broken at the water level because the water ever-so-slightly disrupts the light waves. Light refraction is also one of the reasons why it is difficult to shoot a fish with a bow-and-arrow from a boat as one often has to aim higher than the target to hit it. The depth of the target also forces the shooter to adjust his aim even more.

Figure 2 – 5
Demonstration of light refraction using a pencil and a glass of water

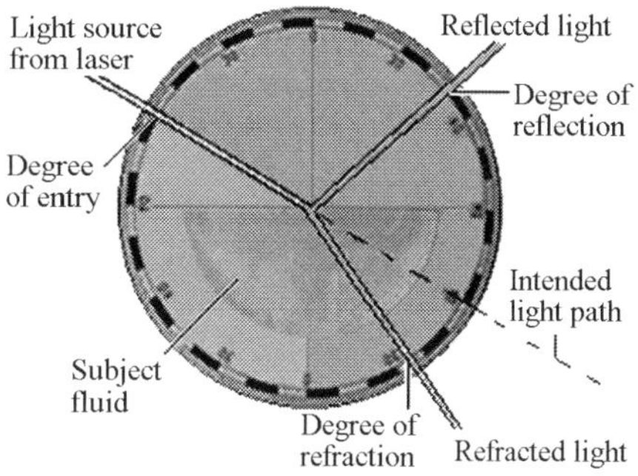

Figure 2 – 6
Demonstration of Refraction & Reflection

Wave reflection and refraction can make it challenging to track, monitor and analyze sound waves when dealing with inconsistent formation composition or multiple subsurface fluid layers. It has also been found that seismic waves are unable to be transmitted through certain types of rock. If an acoustically impenetrable structure is above the target formation, the seismic survey might not show the true value of what is there. Conducting additional trials on the subject property often corrects for these deficiencies

Seismology uses low frequency sound waves. In the early years, dynamite was used to accomplish this task. Today, some still use dynamite, though it is uncommon. Instead, large trucks called thumpers carry very heavy plates that vibrate the ground. These trucks are considered to be better since they do not cause as much damage to the ground, are safer for the work crews, and give off a more precise and consistent wavelength; however, they are limited to flat terrain. A thumper, for example, is unable to hang off cliffs or move in water to analyze offshore projects, and has difficulties moving through dense forests and swamplands.

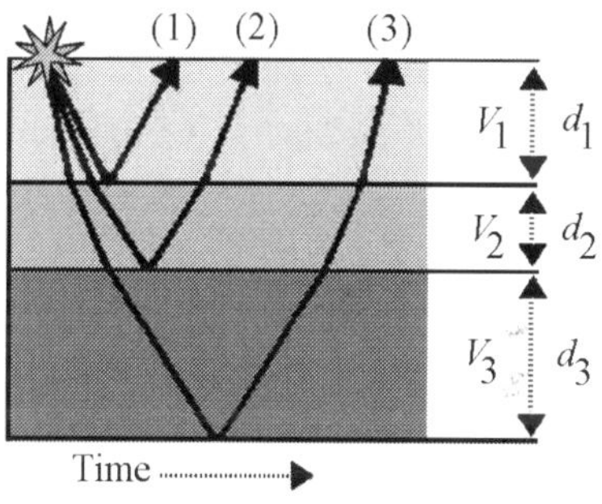

Figure 2 – 7
Seismic Vector Analysis

Recording time in seismic studies is critical to determine the depth of a possible formation. One has to know the elapsed time between dynamite charge (or thumper noise generation) and the time a geophone senses the received sound wave. From there, mathematical formulas involving the traveling sound wave's velocity, time and distance are used to calculate the actual depth of the target. In layman's terms, it can be thought of as vector analysis.

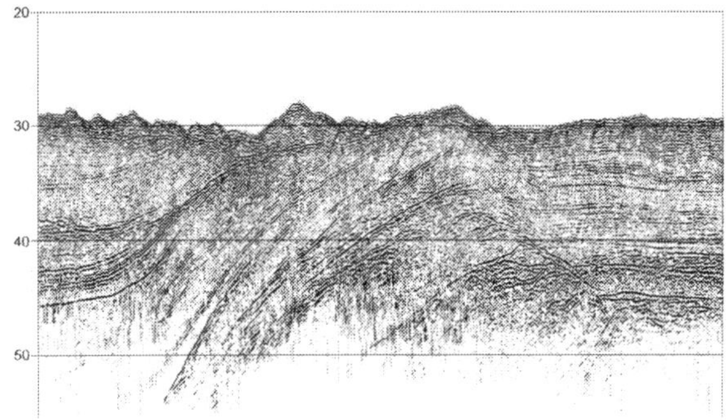

Figure 2 – 8
Sample of a 2D Seismic Report
Source: USGS

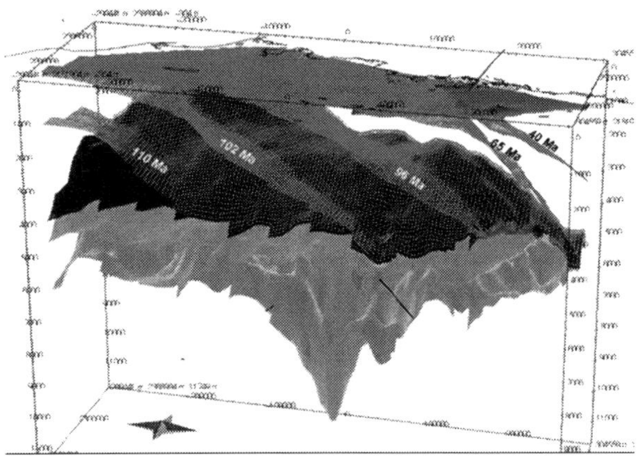

Figure 2 – 9
Sample of a 3D Seismic Report
Source: USGS

The simplest form of seismic analysis, called 2D seismic, utilizes the techniques and methodologies described above. 2D seismic has been used since the 1920s to help isolate commercial oil and gas reservoirs around the world. In the years since, the industry has developed the ability to obtain better estimates of these subsurface structures using what is called 3D seismic.

3D seismic involves the use of supercomputers to arrange multiple sets of 2D data to provide a 3-dimensional image of the target. When combined with an understanding of geology, petroleum scientists are able to tell drilling companies where an oil and gas location may exist.

4D seismic is now possible by conducting 3D surveys over time and combining the resulting data throughout the production phase of a well. Frequently, 4D seismic is used to analyze how efficiently hydrocarbons are being extracted from a reservoir. Offshore seismic is also possible with different equipment, but is currently beyond the scope of this book.

Seismic research has been improving with respect to accuracy. Since sound refraction could be an issue, seismic companies often conduct multiple runs or trials to confirm the results. Often, seismic reports do not show how many attempts were conducted during the study. If only one pass was made, the data might be inaccurate. Conducting multiple trials and altering the distance between geophones often clears up any false assumptions, but also costs more.

Although seismic studies lack the ability to determine the porosity and permeability of a formation, they can provide an explanation of underground structures. The data generated by seismic tends to make more accurate depth estimates than any other survey method.

Exploration Conclusion

As should have been made clear throughout this chapter, exploration is not only complex as to its techniques but also in the reviewing of its resulting data. Remember, it is the people that one should invest in, and not the project as human error can occur when conducting and analyzing these studies. A company whose people not only understand these difficulties but also know what to look for may determine a successful investment.

Regardless of which exploration techniques are used, the goal of this process is to determine if a commercially viable formation exists. If it does, exploration research can tell drillers where to stake the first well, what depth(s) should be targeted, even what types of equipment will be needed.

Unfortunately there is no known survey that can 100% guarantee the presence of oil or natural gas. However, some of these surveys complement each other's deficiencies. For example, 3D seismic can often show subsurface formations but lacks the ability to detect the presence of hydrocarbons. Conversely, radiometric, gravity, and even magnetic surveys may better detect the presence of hydrocarbons, but all lack the ability to distinguish formation depths. They are simply just tools, not the "holy grail" of the industry.

Furthermore, *none* of these surveys are able to detect the porosity or permeability of a formation, one of the reasons why surveys have lengthy disclaimers reminding their customers that the survey does not guarantee *anything*. This makes it difficult to move forward because it is still necessary to drill a test well at some point to prove or disprove a formation. Not all prospects warrant the use of high-tech exploration studies as it may be more economical to drill a test well instead.

Ultimately, these exploration surveys can, at best, help formulate a rough estimate of probable costs necessary to compile an Authorization for Expenditure (AFE) for equipment, such as drilling rigs and pipe. Reserve estimates can be hypothesized to calculate Return of Investment (ROI), which should be used to justify the expenses of further exploration and production. As the saying goes, "high risk yields high reward." Being smart about exploration often mitigates the risk aspect of investing in ventures with tremendous profit potential.

Chapter 3 – Drilling Rigs

Modern-day drilling can be linked back to nearly two thousand years to a basic mining technique developed by the Chinese. The first form of drilling used hand-powered rotary tools to drill a line of small holes into a large rocks slab within a quarry. These holes were later filled with long, dry wooden pegs. After saturating the pegs with water, they would swell and increase in size. This would cause an increase in pressure along the predefined lines, creating cracks that became the outer edges of the smaller block. At first, these blocks would be very rough and jagged along all sides but, after some smoothing and polishing, they would be ready for assembly. This form of mining with the use of hand-held drilling tools is believed to have been used in ancient structures, such as the pyramids of Egypt, many of which are still around today. Moving forward in time, the drilling rigs from Colonel Drake's days were very simple by today's standards.

Cable-tool Drilling Rigs

Cable-tool drilling rigs are, in fact, similar to contemporary rigs to the extent that they share a basic structure and that their power is derived from little more than an elaborate pulley system. Cable-tool rigs were predominately used from the time of Drake's 1859 oil well until the improved rotary method was developed in 1909.

Cable-tool drilling can be pictured by children playing on a teeter-totter. When 2 children are on the teeter-totter their weight is dispersed evenly around the central pivot. But when the child at the end closest to the ground gets off, the child at the elevated end quickly realizes the uneven weight distribution and immediately smashes to the ground.

Figure 3 – 1
Teeter – Totter Example

The walking beam is the pivotal component in a cable-tool rig and operates like the children in the teeter-totter example. One end of the walking beam is attached to a motor, the other to a cable that runs down the borehole to a heavy weight, or series of weights, called sinker bar(s). Interestingly, the first motors used were steam-powered as in Drake's rig. Attached to the sinker bars is a drilling bit shaped like an indented chisel. Each time the walking beam moves up and down the bit slowly smashes its way deeper. This process is very slow. A length of wrapped cable is gradually "let out" of the pulley system as the bit descends. Pulverized rock piles up around the bottom of the borehole which interferes with the interaction of the chisel-bit and the next piece of earth. This debris had to be dealt with to continue the drilling process.

Figure 3 – 2
Cable-tool Chisel Bit
Source: U.S. Geological Survey (USGS)

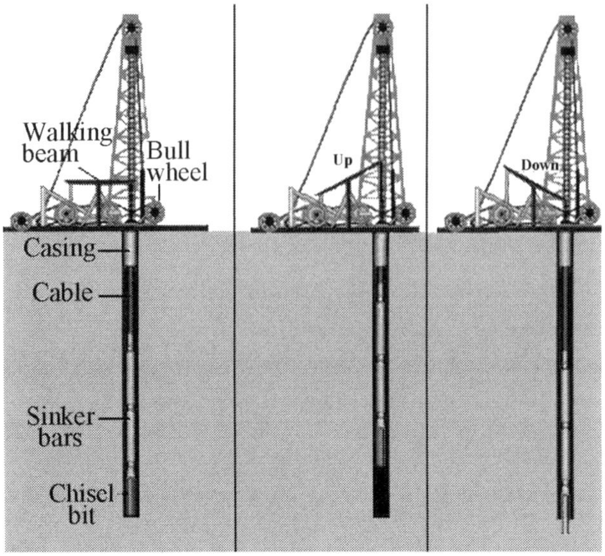

Figure 3 – 3
Components of a Cable-tool Rig

Typically, after each three feet is drilled, the cable is disconnected from the walking beam and reconnected to a bullwheel. The bullwheel is an additional wheel that is located near the ground and passed through the top of the derrick. The bullwheel pulls the cable out of the ground with the use of pulleys. Once the drilling equipment is removed, a special basket called a bailer would be run into the borehole to retrieve and remove the rock cuttings made by the bit. After the bailer removed the cuttings the bit could be reconnected to the walking beam, and drilling would continue. The frequent interruption of having to bail out cuttings is one of the main reasons why rotary rigs are used today. However, some cable-tool rigs are still used to begin a new well, a process called "spudding." Because of this practice, these rigs are now more commonly known as "spudder rigs."

Figure 3 – 4
Example of a Rotary Drilling Rig

Rotary Drilling Rigs

Rotary drilling rigs have been widely used since the 1909 Spindletop well in Texas. Cable-tool rigs were incapable of drilling deep depths and often struggled, or even failed, in formations composed of soft clay or fine sand. Unlike hard rock formation, softer layers are more likely to collapse the borehole. Their cuttings simply can't be bailed out fast enough.

There are several key differences between rotary and cable-tool rigs. Instead of dangling a weighted chisel from a length of heavy-duty cable as seen in cable-tool rigs, a rotary drilling rig uses segmented lengths of hollow pipe ending with a specialized drill bit. This bit utilizes rotational motion combined with downward force produced from the weight of the drill pipe. As the pipe rotates, the bit scrapes and grinds out chunks of rock. Another advantage behind the rotary rig is the drilling fluid circulation system that replaces the need for bailing.

Figure 3 – 5
Example of a Rotary Rig Drill Bit

Drilling Fluid

Drill pipe is hollow to facilitate the movement of drilling fluid to the bottom of the borehole where it is then ejected through nozzles located on the bit. This fluid then flushes cuttings to the surface while improving both bit-earth interaction and drilling speeds. This allows the actual process of drilling to be more continuous. Another function of the fluid is to counteract increasing pressure as the borehole increases in depth. The exact composition of the drilling fluid will thicken throughout the drilling process to prevent borehole collapse.

Size & Specs of Land Rigs

The drilling rig is nothing more than a pulley system. It is limited in its ability to support the weight of the pipe, making it important to consider the depth of the target formation when choosing a rig. A single segment of drilling pipe is usually thirty feet in length and can weigh between 6 and 30 pounds per foot. Accidents can occur if a small rig is used to drill a well deeper than its maximum limit. It is important for not only the crew and staff but also investors and other participants to understand the correct equipment needed to reach the desired depths of the project. The size of a rig is absolutely critical to the success of a well.

	Maximum Depth Limit (feet)
Light-duty	3,000 - 5,000
Medium-duty	5,000 - 10,000
Heavy-duty	10,000 - 16,000
Very Heavy-duty	16,000 - 25,000

Table 3 – 1
Rig Classifications

Offshore Drilling Rigs

Offshore rigs are nothing more than a rotary rig attached to a platform or boat. Different offshore rigs are used at different water and target formation depths.

Figure 3 – 6
Types of Offshore Drilling Rigs
Source: Bureau of Ocean Energy Management

It is interesting to think about how basic the drilling rig really is. Of course, this is not to say that the industry has not made huge advancements in other aspects of oil drilling and production, such as those made in drilling fluid engineering. Despite its simplicity as a concept, the rig itself is still the most important component of the oil and gas industry as it is the foundation of every other action that follows in bringing these commodities to market. Now that the size and types of drilling rigs are better understood, it is now possible to appreciate a discussion of how to put together a project.

Chapter 4 – Putting Together a Project

"Prior Planning Prevents Poor Performance"
-The 5 Ps of Leadership

After geological studies have been compiled and the potential formation depths have been identified, a field development plan is put together for an appraisal well. An Authorization for Expenditure (AFE) itemizes all the costs that would go into the development activities. The costs of drilling include the rig that is to be used along with personnel and supplies, such as pipe and drilling fluid, to name just a few.

With the AFE and field development plan in hand, a proper business plan can be created which could make the project not only economically viable but also lucrative to potential investors. Business plans can be formed in single or multiple phase structures as the field develops. Business and development plans take a long time to put together and many specialized oil and gas experts contribute to them. These plans have to take into account many different factors, such as the creation of roads and infrastructure needed to move and power the equipment. All of the logistics of a project must be predefined in order for it to be successful.

Field Development Plan

Long before the first shovel hits the dirt, the most fundamental aspect of the drilling process is having the legal right to extract hydrocarbons from the subject property. In the industry today, the instrument most often used is a mineral rights lease. With this type of agreement, the owner of the land conveys onto an exploration and production (E&P) company the right to explore and extract minerals, including crude oil and/or natural gas.

A mineral lease agreement usually has a predefined term, five years for example, within which development must be completed or the lease is nullified and the rights revert back to the owner. It is important that every lease has a continuance clause with respect to the term duration. For example, if oil is being produced from the subject property, the lease would continue until production falls below a certain threshold. Commonly, this threshold is set at "the sale of one 'load' of oil per year." If this clause is not in the lease, the E&P company and its investors will not be able to obtain the maximum amount of profits.

Another critical aspect of the lease is having explicitly defined surface access to the subject property for vehicles and equipment. For example, if the business plan does not include surface easements, such as roads and electricity, the project is likely to encounter delays and go over-budget. Such an oversight can cause the need for additional capital from investors. This often cripples new oil and gas ventures.

Having all the details figured out about the property is still another key component of the field development plan. Knowing the exact legal descriptions obtained by a survey is important to avoid possible later disputes with subsurface and surface landowners. The legal descriptions are also necessary for obtaining permits from the state which are needed to ensure environmental and technical standards are fully upheld. For example, advance planning and budgeting should be made for disposal of unwanted substances like salt water. Salt water contaminants are useless to the refinery and expensive to haul away by truck. It is usually more economical to pump the contaminants into a lower subsurface zone with what is called a disposal well where they can be safely disposed of within the earth. These wells require special permits.

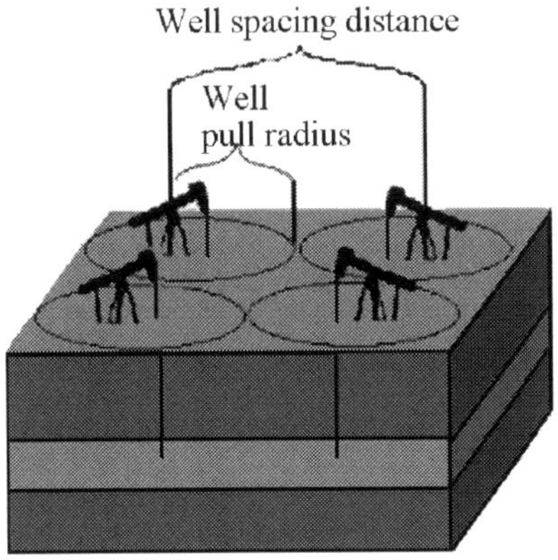

Figure 4 – 1
Well Spacing

Many inexperienced investors believe that one well can adequately drain all of the hydrocarbons from a reservoir. However, due to issues of porosity and permeability discussed previously, a single well can only extract so much from a reservoir. Well spacing is used to optimize the extraction of hydrocarbons, but it is also strictly regulated by the state.

Once a well obtains a permit it is given an arbitrary name and an API number. The API is the American Petroleum Institute, the largest trade organization in the United States oil and gas industry. This organization acts as a self-policing regulatory body, much like the National Association of Realtors in the U.S. real estate industry. Members of the API are expected to adhere to standards of environmental protection, safety, and regulatory compliance above and beyond those simply required by U.S. federal and state laws.

Operating Company

With all the specialists involved in a project, someone needs to be in charge of coordinating and supervising the various independent activities that are contributing to the venture. In the industry this person is called the operator. As the name suggests, the operator is responsible for managing all operations, which means selecting, hiring and paying for the various experts and contractors involved. An operating agreement is an extremely important form for any project. The operating agreement will explicitly outline the daily operating activities that will occur during the various phases of a project.

When hydrocarbons are sold to a refinery, the operator is tasked with disbursing proceeds to whomever they are due. As a result, the operator is also commonly tasked with maintaining all financial records, bookkeeping, and any tax information applicable in whatever state the well is located, typically by hiring a third-party accountant.

Ultimately, it is the operator's primary role to be responsible for compliance with all state and federal rules and regulations. Because of this, the operating agreement is often carefully drafted to best fit the situation and clearly define the expectations of everyone involved in a project.

Typically, the operator is hired by whatever entity owns the right to extract minerals. The operator then has a fiduciary obligation to the owner(s), which means putting their best interests ahead of the operator's in an attempt to achieve the most beneficial results for all involved. Once this relationship is established, the operating company, now acting on behalf of the owner(s), enters into individual service agreements with specific contractors and specialists. This is but one of many different possible ways to structure a venture or project.

Service Agreements

This industry is comprised of many specialists. Often, the company putting together a project and conducting exploration activities will not directly own the drilling and production equipment. Because of this, many agreements need to be put into place before breaking ground actually occurs. Other important agreements involve the various service companies that will contribute their expertise to the project. Before that happens, it is common to "bid out" the tasks that are called for by the business plan. Once the bids have been won, the agreements can be drafted.

When gathering bids from third-party service providers, all of the various aspects and contingencies must be considered and outlined in the proposal. For example, suppose a drilling contractor is selected and given the green light only to get to the field and find the land is still covered with trees. The excavation or clearing of the drilling site was overseen, causing a potential financial disaster. Another example is having an inexperienced operator who tries to clear the drilling pad himself. If the pad is uneven and contains objects, such as exposed tree roots that can damage vehicle tires, there will be a problem of drilling contractors not having proper surface access. Remember: "Prior planning prevents poor performance."

Problems often occur when a specialized group of individuals, such as drillers and roughnecks, arrive at a site where they have to level out the pad or remove tree roots. This often causes two major issues. 1. Drillers get upset at having to conduct tasks that are beyond their expected job requirements. 2. It tends to cause financial issues with respect to the AFE. The reason is simple: paying a specialized company for excavating is generally cheaper than paying a much higher rate to drillers who will be less efficient at the task of clearing. It is not wise to pay a CPA to paint a house. It is questionable if one should even want their CPA to be their financial advisor. The job functions are different and require unique skill-sets and tools.

A tremendous amount of work and planning must be competently performed when creating these business and development plans. Forgetting to plan for any contingency will probably lead to disaster later on as it is difficult to change course after drilling has begun. This next chapter will focus on the equipment involved in a typical land-based development project using a rotary drilling rig.

Chapter 5 – Drilling Site Preparation

Once the business and development plans have been finalized and contractors have been hired, a project can start to break ground with excavation and site preparation. Some of the first people on the drilling location are the excavators. They clear trees, level surfaces for the rig, and dig pits. Many times they will put in temporary or permanent roads to facilitate the arrival of various pieces of heavy equipment.

Once the site is cleared, excavators prepare special areas known as drilling pads and layout yards. They typically bring in heavy equipment like bulldozers and other earth movers. The individual in charge of the excavation process is usually called the clearing foreman. The clearing foreman oversees all excavation personnel, ensures compliance with OSHA safety guidelines, and meets all environmental requirements. The basic layout of a rig site can be configured in many different ways, but a basic example is depicted in Figure 5 – 1.

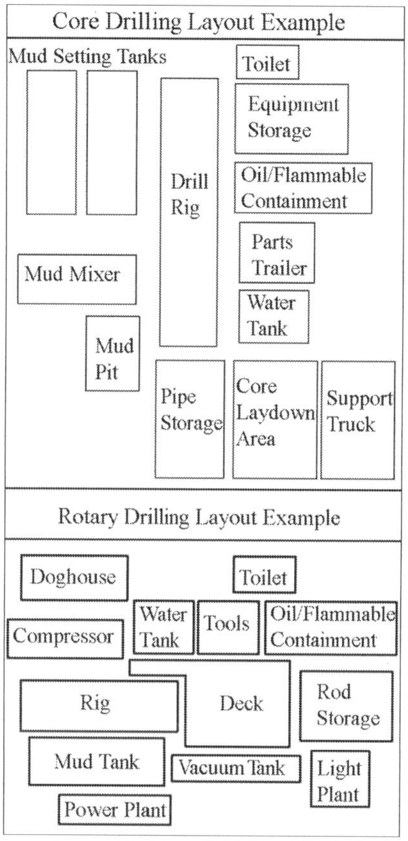

Figure 5 – 1
Drilling Layouts
Source: U.S. Forest Service

The drilling pad and associated roads are critical for the excavation phase. The pad must be level to prevent serious problems during drilling. On loose and uneven terrain, a pad made of dense material, such as cement, is often used. Aside from the need to be level, the pad must also be durable. It has to be able to support the derrick itself which, in turn, supports thousands of pounds of drill pipe both above and below the surface. It should be noted that horizontal drilling is not accomplished by using an unlevel foundation.

Larger projects sometimes have a cellar, which looks like a cement pad but functions somewhat differently. Located underneath a rig, a cellar is typically made of thick concrete and provides additional space between the rig floor and the ground. This is where parts that would otherwise have to be located below ground can be installed.

Excavating companies may also dig a shallow pit, called a "water pit," to store freshwater which will later be used to make drilling fluid. The water pit is usually just a widely dug out hole that holds water brought in by truck or pumped in by a nearby water well. Running long water lines from a water source, such as a pond, lake or river, to the drilling location may be more economical than drilling a brand new water well or having it trucked in. Oftentimes, water pits will be lined with plastic if oil- or saltwater-based drilling fluids are to be used in order to prevent soil contamination. Nets and fences are also used to keep out birds or livestock.

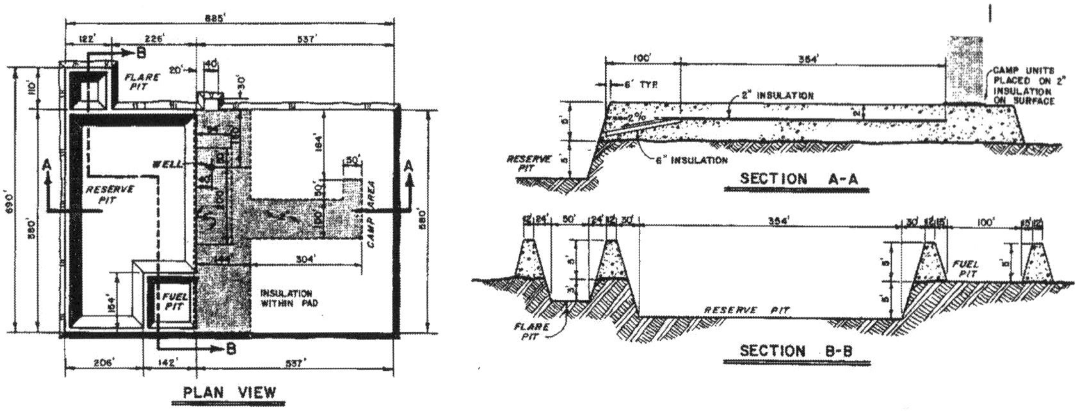

Figure 5 – 2
Detailed Pit Diagrams
Source: U.S. Geological Survey (USGS)

The water storage and drilling fluid mixing areas can use above ground tanks as a more efficient alternative to these pits. There can also be a pit or series of pits to collect dense cuttings, like shale, which is later moved to the reservoir or flow-over pits. A channel is dug between the borehole to a special cuttings collection pit from which geological samples are taken and analyzed.

Figure 5 – 3
Mud Pumps

Once the site has been prepped, heavy equipment can be brought onto the location. Many trucks are involved in moving equipment to the site which will later be moved into place by cranes and bulldozers. The configuration of the equipment as it arrives on site depends on the size of the rig and the intended target depth. Smaller rigs designed for shallower depths can be transported on-site fully assembled with one single truck. Medium- and heavy-duty rigs are disassembled at the end of each use, transported via several vehicles (even helicopters and planes), and assembled on site. This modular design allows a rig to service multiple locations as quickly as possible.

A series of heavy-duty mechanical pumps and engines are trucked in and set up near the mixing areas. One of the most important of these is the mud pump, which facilitates the drilling fluid circulation system by keeping a flow of pressurized mud through the drill string. Industry workers use the terms "mud" or "drilling mud" interchangeably for drilling fluid because the fluid used in the early days of rotary rig technology was literally nothing more than wet dirt. The role of drilling fluid in the actual drilling process will be thoroughly described in Chapter 7 on Drilling.

The movement and assembly of heavy equipment is one of the most exciting times of a new project as years of preparation and planning are carried out. The only feelings that top this are actually hitting oil and the nail-biting anxiety of getting closer to a targeted formation.

Once all the equipment is brought to the site, an apparatus called a structure is assembled directly over the location where the well is to be drilled. It provides a level work area for the crew, called the rig floor. Between the rig floor and the ground, a special device composed of a series of valves, called a blowout preventer (BOP), will be installed after the surface casing is set. A BOP prevents the type of oil explosion made famous by the Spindletop well. The eruption of oil from that well, while exciting to watch, had negative implications for the environment and made many barrels of oil unrecoverable. More recently, the fire observed in the Deepwater Horizon well disaster in the Gulf of Mexico in April 2010 was thought to be associated with a faulty BOP.

Figure 5 – 4
Rig Structure

A doghouse is another modular structure that is usually located at the rig floor level if not close by. It is similar to a portable trailer and contains an office for the crew as well as tool storage. Third-party specialists, including the mudlogger and company man, maintain their own offices in other portable buildings located somewhat near the drilling location, though not necessarily on the rig floor itself.

Figure 5 – 5
Blowout Preventer (BOP)

The most visible and recognizable component of the drilling rig is the derrick. A derrick is a tall steel structure that supports the weight of the heavy drill pipe, bit, and circulating drilling fluids. With the drilling pad prepared and the rig on site, the drilling supervisor will double check the rig assembly and hold a safety meeting with his crew that discusses the various components of the rig, the crew's responsibilities, and the project's operational guidelines as a whole.

Figure 5 – 6
Derrick of a Drilling Rig
Source: Bureau of Land Management (BLM)

Chapter 6 – Components of a Drilling Rig

Drilling a well requires two major forces: lift and rotation. These components also need a power source to function during actual use. However, the role of power as it applies to these two forces will be saved for the next chapter. This chapter can be thought of as "drilling theory," while the next chapter will see these concepts applied in practice. Each piece of equipment is important for the role it plays in either lift or rotation.

Lift

The concept of a derrick is similar to scaffolding. A scaffold is a sturdy, temporary structure that supports the construction of another structure, using as few materials as possible. The height of a derrick is related to the intended target depth a rig is expected to reach. The deeper the intended drilling activity, the more drill pipe required, which increases the total weight the derrick needs to support. In other words, the derrick, supporting the weight of all other components, is the foundation of the rig lift system.

Figure 6 – 1
Rig and Drawworks

The drawworks is an essential piece of equipment that houses a spool around which crew members wrap steel cables called drilling line. The rig's lift system is composed of the structure, derrick, and the drawworks, with drilling line connecting the parts. The drawworks not only moves equipment around the rig floor, but also lifts the pipe into and out of the borehole.

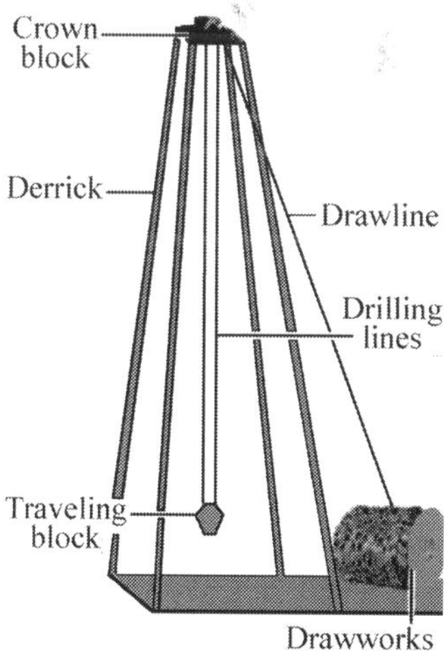

Figure 6 – 2
Pulley System with Drawworks

The drilling line is a length of braided steel cable wrapped around the various pulleys, which are called sheaves. These particular sheaves are located at the very top of the rig in what is called the crown block. The stationary crown block rests on top of the derrick to support the tremendous weight of the drill string used throughout drilling.

The drilling line continues from the drawworks to the crown block into another set of sheaves called the traveling block, which supports the weight of a rotating mechanism called the kelly. The drilling line is wrapped between these sets of sheaves multiple times, allowing the drawworks to lift and lower many thousands of pounds of weight with just a fraction of the force that would otherwise be needed. The traveling block is alternatively raised and lowered by the drawworks.

The rig's lift system is based on simple concept of a pulley, as more wheels (or sheaves) are added to a crown block or traveling block, less force is needed from the drawworks. In other words, a smaller motor can be used as one adds more wheels and cable to the pulley system. With that said, it is important to understand that the pulley system of a drilling rig is limited by the rating of the derrick height and weight limitation specifications, not the size of the drawworks motor.

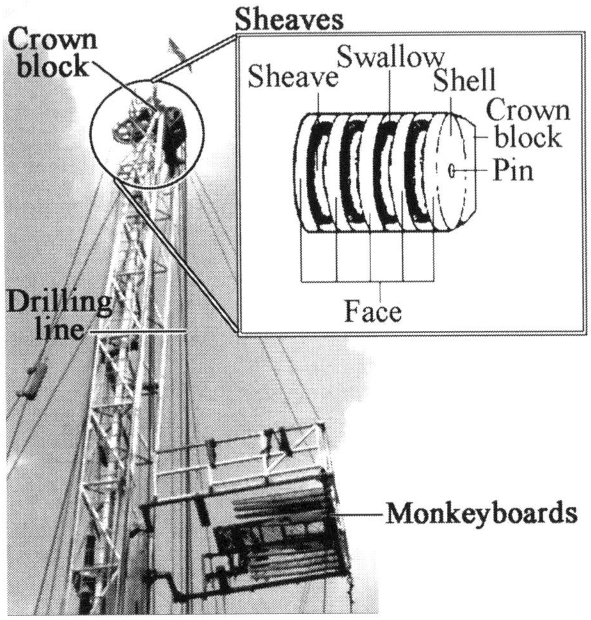

Figure 6 – 3
Sheaves, Line, Crown Block and Monkeyboards

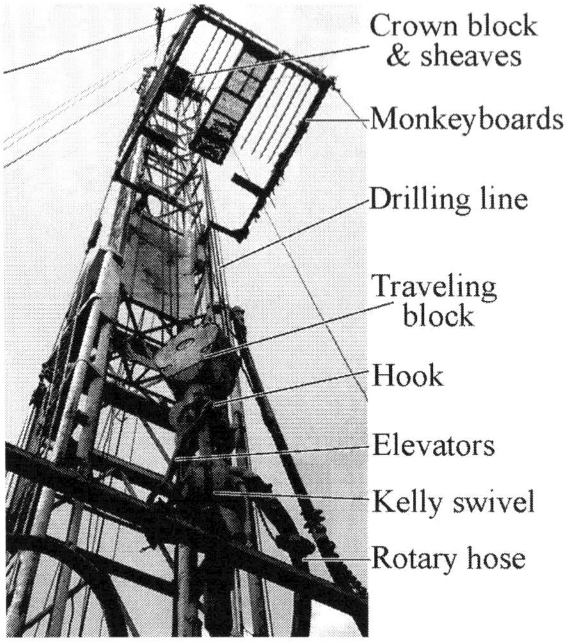

Figure 6 – 4
Components at the Top of the Derrick

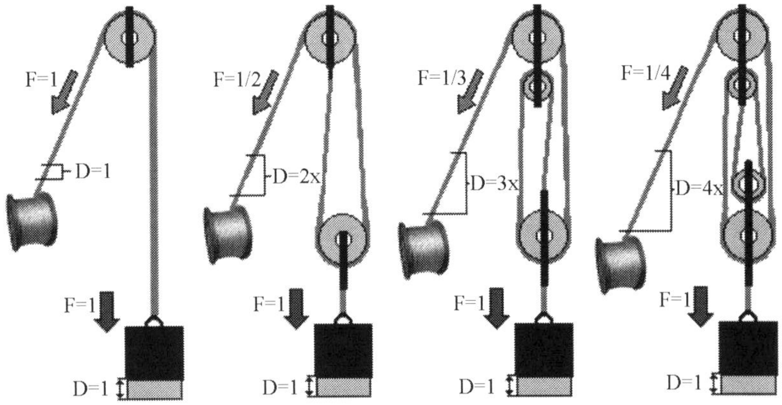

Figure 6 – 5
Basic Concept of Pulleys and Lift Forces

Catheads are auxiliary winches powered by the drawworks used for various functions in the drilling process. One of their main functions is to mechanically operate the tongs when connecting or disconnecting drill pipe segments. In addition to the primary lift system, an air-powered hoist is usually present to help perform other tasks like moving the next segment into position.

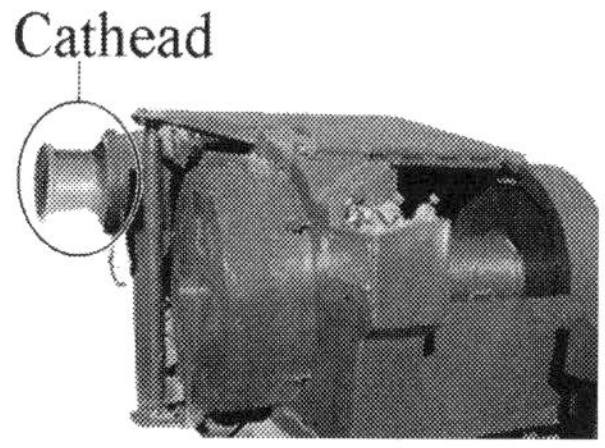

Figure 6 – 6
Cathead of a Drawworks

The final component of a rig lift system is the hook at the bottom-end of the traveling block. When drilling, the hook connects the rig's rotation and lift systems with a swivel. Because of the swivel, the hook does not rotate, meaning neither the traveling block nor any other lift component ever rotate.

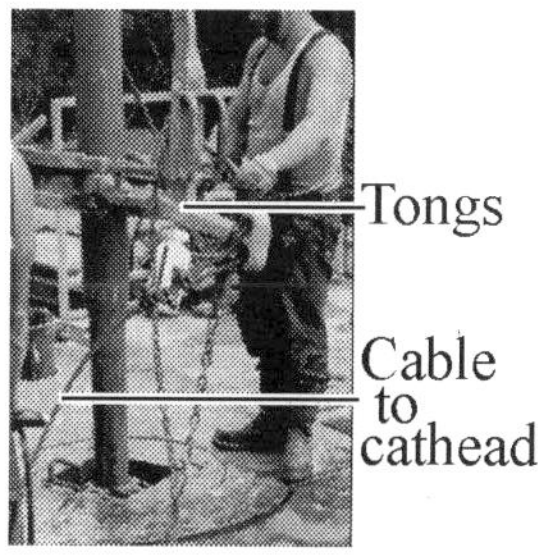

Figure 6 – 7
Rig Hand with Tongs

Rotation

A special type of pipe segment called a kelly is attached to the swivel. The kelly is not shaped like normal pipe; instead of being circular, the kelly is either square or hexagonal. The kelly is then connected to the rotary turntable through a mechanism called a master bushing. Located on the rig floor, the turntable powers the rotation of the kelly-bushing system. The drill string, consisting of many segments of heavy pipe and ending with a bit, is then attached to the bottom of the kelly. Figure 6 – 9 is a simple picture of the components that interact with the kelly.

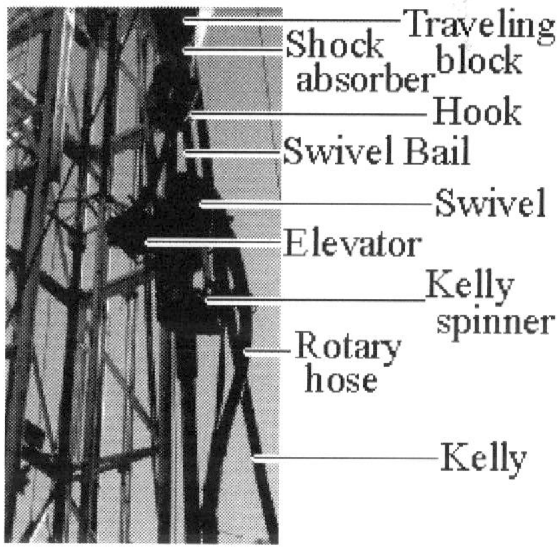

Figure 6 – 8
Intersection of a Rig's Lift & Rotation Systems

While the turntable system is not the only rotary drilling technique around, it is by far the most common. Another notable system is the top drive. Also referred to as the power swivel, this system uses a motorized swivel which hangs from the traveling block. This device allows for the replacement of the kelly-bushing apparatus as the drill string is directly connected to the traveling block. This results in an increased length of drill pipe that can be drilled into the borehole before the next segment of pipe needs to be added, thereby increasing efficiency.

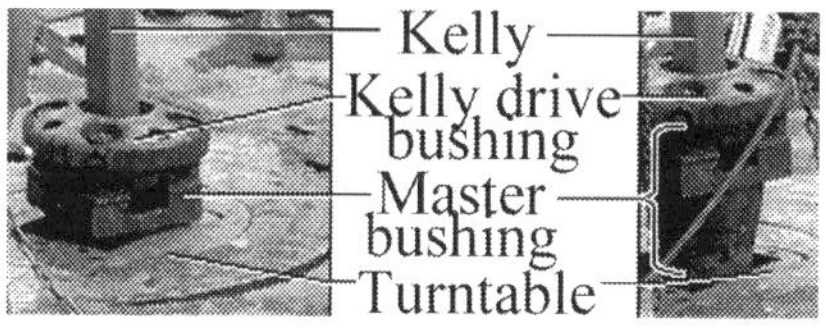

Figure 6 – 9
Rig's Rotary System: Kelly, Kelly Drive Bushing, Master Bushing, and Turntable

Issues with either spin or lift are frequently the cause of many problems when drilling. Being aware of these two forces will help a novice better understand the equipment used. If problems occur while drilling, they can typically be related back to one of the components mentioned in this chapter being missing, damaged, or improperly used. A good example would be a broken bit. If a drill bit breaks, the rotational forces will not be as effective. This will be better demonstrated throughout the next Chapter on drilling.

Chapter 7 – The Drilling Phase

With the components and concepts of the rotary drilling rig defined, it is now possible to appreciate how these forces come together when actually drilling. There are many ways to drill a hole in the ground, but it is vitally important to understand how to avoid costly mistakes that often bankrupt many new exploration and production companies. It is only after being personally involved in numerous projects that I am able to adequately explain the details of these concepts in an easy to comprehend manner for those who are less experienced.

This book interchangeably uses the terms borehole, wellbore, and oil well to refer to a hole and the process of creating that hole. There are very subtle differences between these terms, but for the remainder of this book the reader could consider them the same thing.

Borehole: a vertical or horizontal hole in the ground. This term may or may not refer to oil or gas.

Wellbore: is any hole drilled for the purpose of exploration and extraction of natural resources such as oil, natural gas, or water, if present.

Oil/Gas well: a hole bored through the earth's surface to find or extract hydrocarbons, such as oil and gas.

Drill Bits

One of the most important components of the drilling process is the bit. Rotary drilling bits allow for different types of cutting through both a wider variety of formation compositions and greater depths than older chisel-style bits. A rotary rig drill bit is typically classified as being either roller cone or fixed as seen in Figure 7 – 1.

Figure 7 – 1
Types of Rotary Drilling Bits

Roller Cone

Roller cone bits have steel cones, usually three, with sharp steel teeth or small, blunt, tungsten-carbine inserts that cut and grind as the bit rotates. A series of nozzles protrude from the drill bit that sprays high-pressure drill fluid to cool the teeth and to continuously flush away the various cuttings created by the bit. By clearing away these cuttings, the drill bit is exposed to new rock, greatly increasing its efficiency.

Figure 7 – 2
Roller Cone Bit with Steel Teeth

Roller cone bits often do not have their own motors but are instead turned by the rotary system above. As the bit rotates clockwise, its cones move independently due to a system of ball bearings. When the rotation of the drill bit is combined with the weight of the drill string, the bit is able to grind and cut away rock that is in continuous contact with its teeth or inserts. However, whenever one of these bearings wears out, the cone it is attached to will jam and could possibly break off from the bit if not detected. Though rare, if a cone has fallen off the bit, a costly fishing process must take place as a broken cone at the bottom of the borehole may impede further drilling efforts.

Figure 7 – 3
Roller Cone Bit with Inserts

In addition to the potential failure of ball bearings, the teeth or inserts on the end of a bit are gradually worn away over time. It is not uncommon to have a tooth or insert break while drilling. Unlike the major disruption caused by broken cones, a broken tooth or insert only moderately slows the drilling process as these smaller pieces are often flushed to the surface with the rock cuttings. As more teeth or inserts from the bit are broken off, the ball bearings of the individual cones will tend to wear more quickly due to an uneven interaction between bit and earth. Rotary drill bits are expected to break or wear down over time.

Fixed

A fixed rotary drill bit does not have any bearings or moving parts. They can be thought of as blunt objects with a unique configuration of sharpened or jagged edges. As the drill string rotates, the bit's edges drag against rock at the bottom of the borehole and break down the formation into rock cuttings. As with roller cone bits, the nozzles of the fixed bit inject drilling fluid to flush cuttings up to the surface. Fixed bits are typically made of steel, natural diamond, or polycrystalline diamond compact (PDC).

Figure 7 – 4
Fluid Nozzles

Whether to use a roller cone or fixed bit depends on the intended target depth, the type of expected rock formations, and the overall economics of the project. Fixed bits are considerably more expensive, but can be more efficient in certain situations.

Spudding the Mousehole and Rathole

Spudding is an industry nickname for beginning a new borehole. After excavating the surrounding areas, erecting the derrick and connecting the related equipment, it is time to start the actual drilling phase by spudding various holes. The first two are the mousehole and the rathole located a short distance away from the planned borehole.

The mousehole in Figure 7 – 5 is where the next segment of pipe is held before it is added to the drill string. Notice the close proximity of this hole with the kelly-bushing and rotary turntable.

Figure 7 – 5
New Pipe Waiting in Mousehole

The rathole is a hole in the rig floor and ground, also located near the borehole. It holds the kelly, bushing, swivel, and rotary hose unit; this allows the traveling block to be free for lifting tasks associated with adding and removing pipe segments.

The important thing to remember about the mousehole and rathole is that they are simply auxiliary, shallow holes of less than 30 feet, the length of a single pipe segment. These two holes aid the crew throughout the entire drilling process. With the rathole and mousehole prepared, the actual borehole is spudded.

Figure 7 – 6
Rathole and Kelly

Drilling Fluid Circulation System

The role of drilling fluid begins at the nozzles located on the drill bit at the bottom of the borehole. As the fluid passes through the nozzles, it cools the bit. After being ejected by the nozzles, the fluid washes away cuttings from the bottom, allowing the bit to contact the next piece of earth. On its way back to the surface, and under a great deal of pressure, the fluid coats the walls of the well, providing stability and helping to prevent collapse. Finally, when it reaches the surface, the cuttings are analyzed by a team of professionals to determine what types of formations and materials the drill bit is encountering.

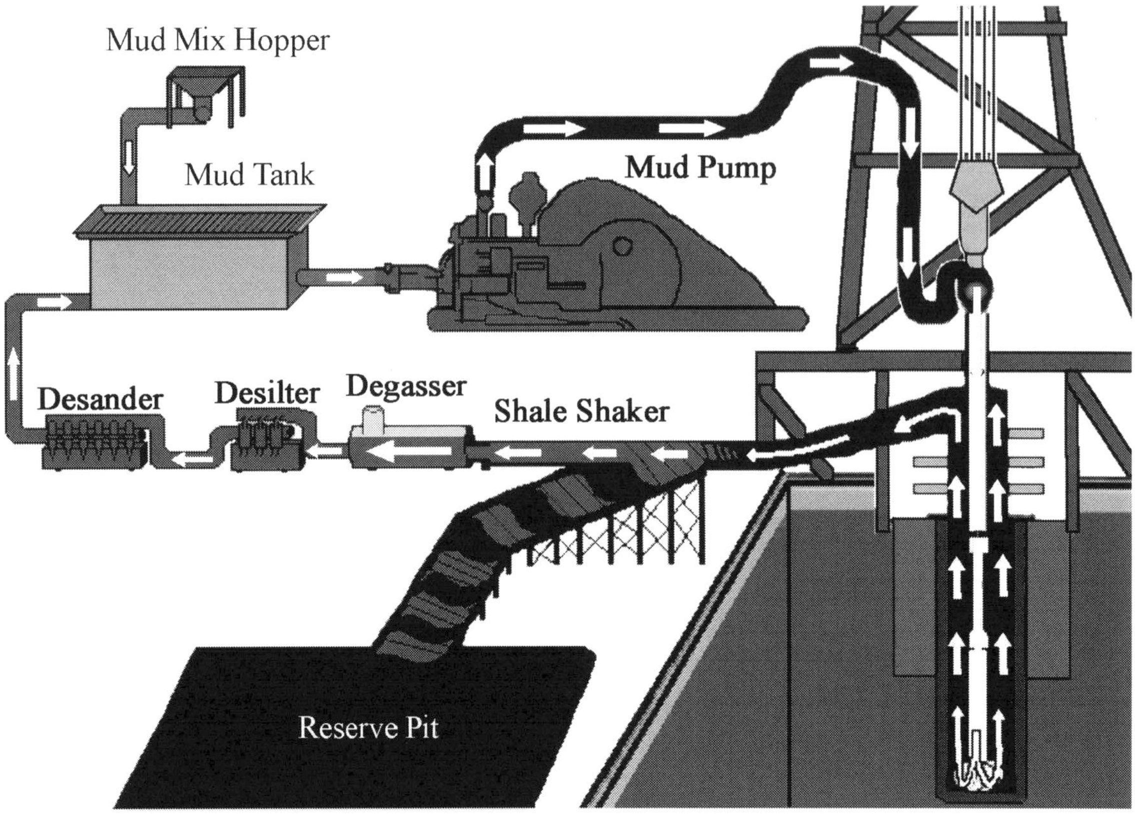

Figure 7 – 7
Fluid Circulation System Components

Figure 7 – 8
Hoghead Checking Mud Consistency

The fluid is created at the mixing hopper station and mud pit. Its composition will vary throughout the entire drilling process due to the forces of pressure, which increase as drilling continues. The pit is connected to a mud pump with a suction line.

Meanwhile, at the bottom of the borehole, the bit grinds and cuts out small pieces of rock. The fluid comes out under so much pressure and speed that it dislodges any large pieces of rock that might be stuck. Next, the fluid carries these rock cuttings up the empty space, called the annulus, between the drill string and the wall of the well.

At the surface, the fluid and cuttings are separated: some cuttings are removed and sent for analysis, while the fluid is filtered of impurities by several machines, such as a shale shaker, a desander, and a centrifuge. The vast remaining bulk of the cuttings are disposed of in tanks or pits. Sensors are also used at the surface to detect gas throughout the drilling phase. These gasses are not only recorded but analyzed by a mudlogger. A mudlogger may recommend additional mud be added if downhole pressures suddenly increase.

Mud is also added to compensate for any loss of fluid due to buildup along the well walls, or that may have been expelled as cutting waste. The buildup of mud along the walls of the well, referred to as "cake," helps maintain the structure of the borehole as pressure and depth increase. A mud engineer, sometimes called a hoghead, supervises the physical change in mud consistency and regularly adjusts mud composition according to density, pressure, and formation conditions.

The drilling circulation system can be related to the human circulation system: blood is created in the bone marrow (mud mixing station and pits), moves through the heart (mud pump), and through arteries (rotary hose and drill string) to wherever it's needed. On its return trip back through the body's veins (the annulus of the well), drilling fluid is comparable to de-oxygenated blood bringing waste materials to filtration systems like the spleen, liver, and kidneys (shaker, centrifuge and desanders). Just as blood is continuously filtered and reused, the now-clean drilling fluid goes back to the mud pump for recirculation.

A healthy circulatory system is critical to our lives. If an organ in the human body is faulty, or damaged by misuse, it can lead to death. The fluid system used during drilling is the same: equipment that is faulty, not properly looked after, or misused by the drilling crew can lead to financial disaster.

About the Rig Crew

It can be argued that the single most important decision an operator will make is selecting the correct contractors. The operator and drilling company will sign a contract according to either a day rate, a depth-of-drilling (also called a footage) rate, or what is referred to as a "turn-key" rate. Day rates and footage rates are fairly straightforward, but a turn-key rate can be somewhat complex because the driller has to estimate, in advance, all of the drilling costs. There are many unknown factors and unplanned events which might occur due to no one's fault. An unplanned event could easily throw a project over-budget. This is why many projects set aside money for what is referred to as "contingency."

The individual in charge of the rig crew is called the rig superintendent. Often nicknamed toolpusher by his crew, the rig superintendent oversees drilling operations, ensures crew safety policies, and furnishes the operator with regular status reports. This cooperation is important because the operator needs to know how far along the drilling process is in order to bring in the next group of experts, such as cementing and casing crews.

Many times there will be an assistant rig superintendent, called a night toolpusher. As the name suggests, this person aides the toolpusher during the day and/or oversees all drilling tasks performed during the night shifts.

When a large field is being developed with multiple rigs, or when a globe-spanning E&P company needs help organizing multiple geographic areas, an area drilling superintendent is brought in to coordinate the various activities between sites or regions. This person's responsibilities are similar to a district manager for a major quick-service restaurant chain. In the restaurant business, a district manager spends most of his or her time traveling between stores and making sure each is compliant with the regulations and policies of the company. Likewise, an area drilling superintendent makes sure each rig has what it needs to complete its task on time and on budget.

Figure 7 – 9
Top of Derrick and Monkeyboards

Figure 7 – 10
Usage of Slips and Bowl

Each rig has a driller working for the toolpusher. The driller, or drilling foreman as he may be called, is not only in charge of supervising the rig's floor crew, but also operates mechanical equipment like the drawworks and rotary table. The driller is responsible for ensuring that equipment is properly maintained and correctly used as many of these components are extremely expensive to replace if damaged.

The actual crew working on the rig floor goes by many names: floor hands, rig crew, deckhands, but the name by which most of these individuals pride themselves is roughnecks. Drilling for oil is a rugged, time-consuming, and difficult process. It also used to be extremely dangerous, though advanced techniques and new safety measures have gone a long way to reduce risk.

Even among these roughnecks there are specialists. The roughneck who works at the top of the derrick during the tripping pipe process, on what is known as the monkeyboard, is called the derrickman. When the crew is not tripping pipe, he works with the mud engineer by adding whatever chemicals or additives are deemed necessary according to current drilling conditions. The derrickman also maintains the machinery and components of the derrick.

A slip puller is the roughneck in charge of operating heavy-duty tools called slips. Slips are metal wedges that are placed between the drill string and the rotary table. These slips are shown in Figure 7 – 10. They hold the weight of the drill string when it's time to add or remove a segment of pipe. Depending on the size of the project, roughnecks may be assisted by a team of junior crewmen, called roustabouts, who perform supplemental tasks, such as moving equipment around the rig floor and maintaining tools. Roustabouts may also be referred to as worms by their more seasoned colleagues.

A representative from the operating company who remains physically at the drilling location is called the company man. For smaller, independent oil companies, the company man will simply be the CEO of the operating company. It is vitally important for a company man to be on the scene to coordinate the activities of the various contractors and service companies that need to be brought in at appropriate times. A participant in a project who observes – or is told – that a representative of the operating company is not on-site during drilling should consider this a red flag that the operator may be up to some unscrupulous activity, may not be taking the job seriously, or may simply be too incompetent to appreciate his role in the project.

Third-party supply companies are sometimes brought in if extra equipment is needed, such as drill bits, pipe, fluid additives or even meals for the crew. It's important to make sure these types of expenses are planned for in the A.F.E.

Doing business in America today requires an almost burdensome amount of office-administrative support for such purposes as taxes, managing payroll, filing (and documenting) invoices, ensuring compliance with federal, state and county agencies, public relations, and so on. The important thing to remember is that these services are required and should not be seen as "a possible way to trim overhead" by cutting corners on a project's budget. For example, some E&P companies have a full time petroleum geologist on staff, though others will have one or more available on an on-call basis. Many of these office expenses are simply a cost of doing business, but a properly run office support team can go a long way toward making the actual in-the-field drilling process a resounding success.

Drilling Collars and Pipe

The first few segments to be connected to the drill bit are called drilling collars. Collars are used to add more force (weight) to the drill bit. A single drilling collar segment is usually 30 feet in length, weighs more than a ton, and ranges in diameter according to the specifications of the intended borehole. Drilling collars and drilling pipe are both hollow to facilitate the flow of drilling fluid. They are connected to each other by what is known as a pin and box. A pin can be thought of as the male end, while the box can be thought of as the female end of each pipe segment, as seen in Figure 7 – 12.

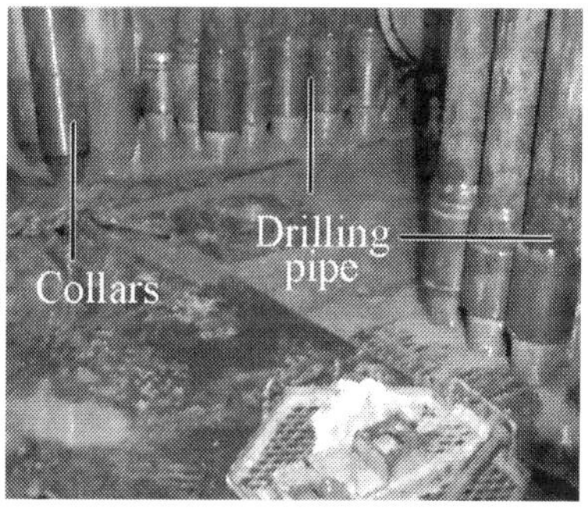

Figure 7 – 11
Difference between Drill Pipe and Collars

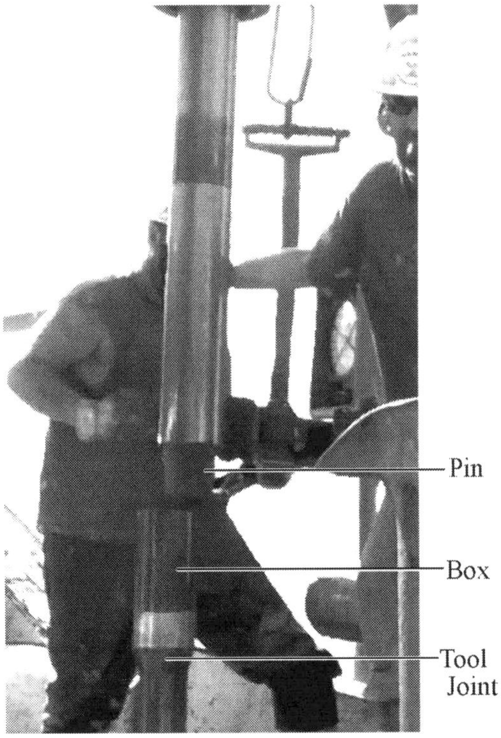

Figure 7 – 12
Tool Joint Connectors: Pin and Box

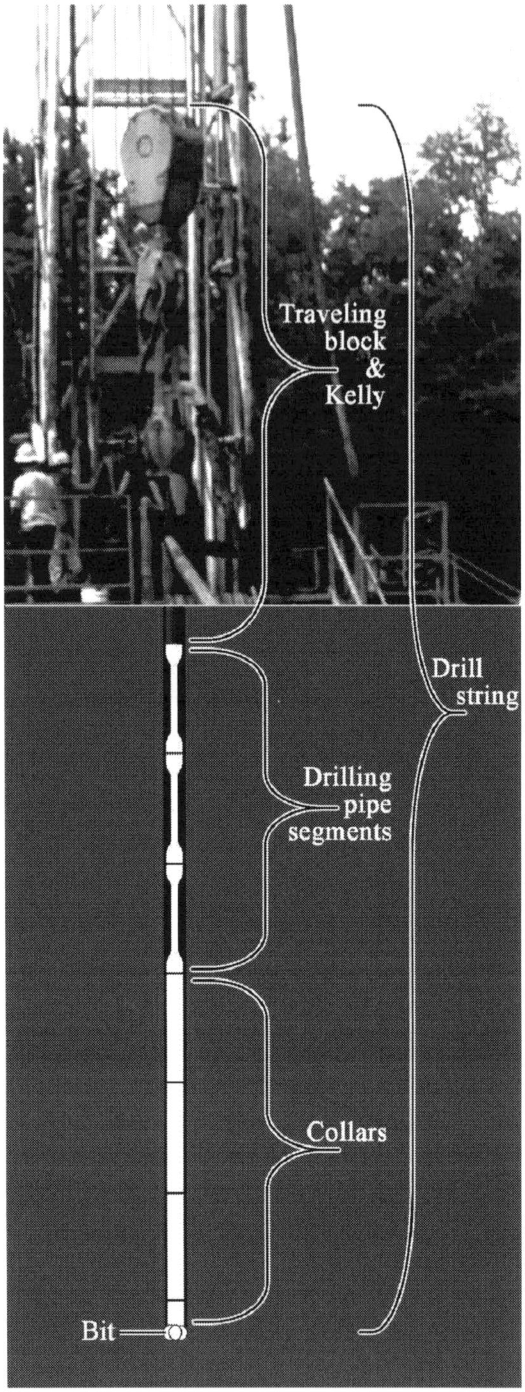

Figure 7 – 13
Drill String Components

Drilling collars are cylindrical and heavy as they are designed to weigh more than drilling pipe. They are made by extruding hot steel through a molded nozzle to create a hollow tube with the desired pipe wall thickness. Once the drilling collar tubes are created, the pin and box ends are machined on in a process like sharpening a pencil on one end, along with a process similar to core-threading on the other.

When drilling, the weight must be on the very bottom of the borehole, as close to the bit as possible. This is accomplished by creating pipe that is shaped differently than collars. Drill pipe is meant to be strong but lightweight, a challenging concept because it is currently impossible to make something weigh nothing and be indestructible. Therefore, drill pipe is created by extruding a steel tube similar to collars, but of smaller diameter. Instead of machining the ends of the drilling pipe, prefabricated threaded steel pins and boxes are welded. These prefabricated pin and box ends are known as tool joints. Since the tool joints are welded instead of being machined directly into the pipe, the drill pipe tends to be weaker at these joints. On the other hand, collars need to be more durable as they are subjected to the deepest depths and the highest pressures and temperatures, while simultaneously acting as the foundation for the entire drilling apparatus. A drill string is composed of the drill pipe and collars, and can be seen in Figure 7 – 13.

Adding to the drill string is accomplished through a labor-intensive process called "making a connection," which involves the use of much of the rig's equipment. This process begins with Figure 7 – 14(A): each time the kelly reaches the rig floor, an additional segment of pipe needs to be added. In Figure 7 – 14(B), the driller uses the drawworks to raise the full length of the kelly and part of the last pipe segment a few feet above the rig floor. In Figure 7 – 14(C), the slip puller secures the top of the pipe to the rig floor with the slips, then mechanical tongs disconnect the kelly from the last segment of pipe. In Figure 7 – 14(D), the kelly then moves above and connects to the new segment of pipe located in the mousehole. In Figure 7 – 14(E), the kelly and new segment move above the existing drill string, still attached to the rig floor with slips. Rig hands will use tongs to make these connections.

Once the kelly and new segment of pipe are reconnected to the drill string, the driller raises the traveling block a few feet off the ground to release tension from the slips so they can be removed, thereby putting the weight of the drill string back on the derrick. In Figure 7 – 14(F), the drill sting is lowered back down until the kelly bushings fit into the turntable of the rig floor. Finally, in Figure 7 – 14(G), an air hoist moves the next new segment of pipe from the pipe rack into the mousehole to await the next connection. This process will repeat itself approximately every 30 feet, the length of each drill pipe segment.

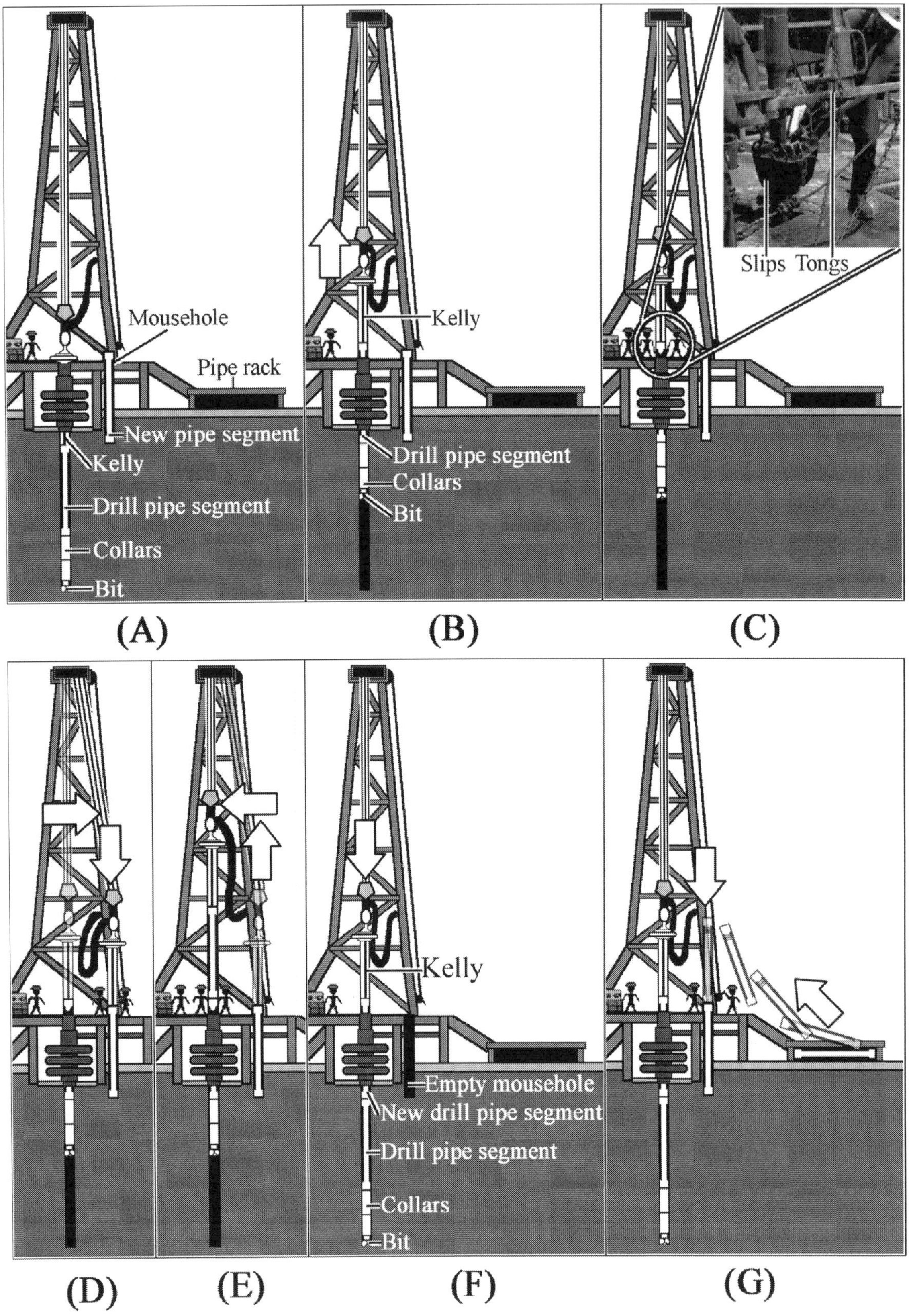

Figure 7 – 14
Making a New Pipe Connection

Tripping Pipe: Removing the Drill String

Time is money, and when the call is made to remove the entire drill string, a process known as tripping pipe is undertaken. This process is similar to making new pipe connections, only in reverse. First, the slips are inserted between the drill string and the rig floor. The kelly swivel and bushing unit are then detached from the traveling block's hook and stored in the rathole. Once the traveling block is free, it can lift several pipe segments above the rig floor with the use of elevators located on its side.

Figure 7 – 15
Disconnection of Kelly from Traveling Block

It is usually a good idea to disconnect several segments of drill pipe together to save time, usually between two and four based on the height of the derrick. Now called a stand, these combined pipe segments are then disconnected from the drill string and stored in a vertical position, held in place at the bottom by what is known as the rack, between the rig floor and monkeyboard.

There are two important tasks performed while placing the stands in the rack. The driller and roughnecks at the rig floor work together to guide the bottom of the stands into the correct position of the rack. The derrickman, who is at the top of the monkeyboard, disconnects the elevators from the stand and guides them into the appropriate grooves, or fingers, of the monkeyboard. While doing this task, the derrickman is attached to a safety harness that is connected to the rig. This is repeated over and over again until the entire pipe is removed from the hole. Later, this procedure is reversed to place the pipe back into borehole.

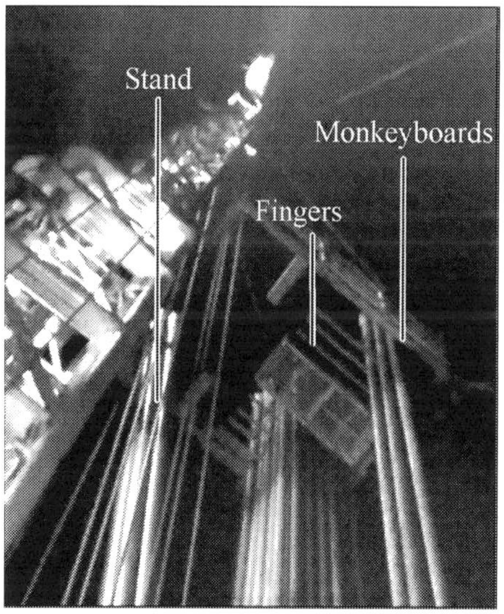

Figure 7 – 16
Stand Entering the Monkeyboards

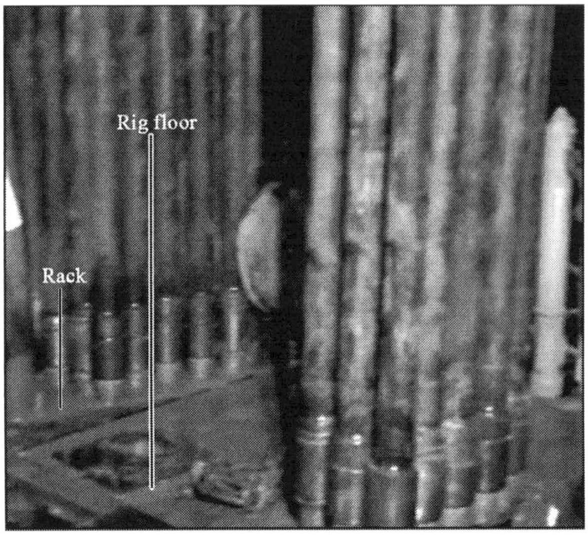

Figure 7 – 17
Pipe Stands Entering Rack on Rig Floor

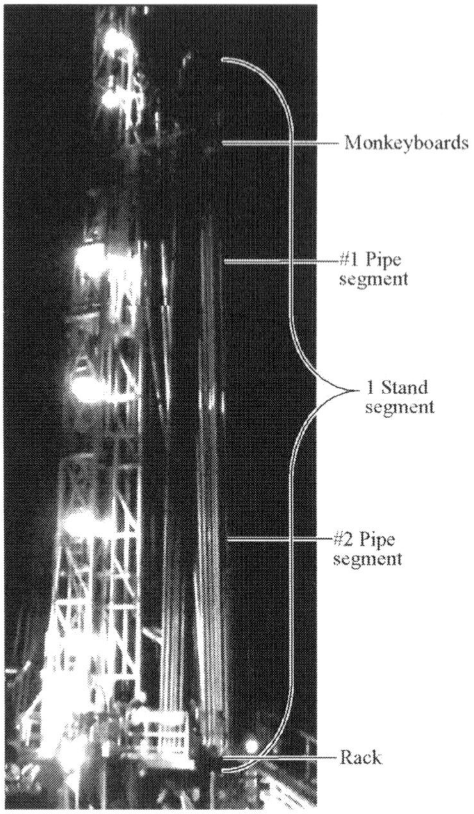

Figure 7 – 18
Example of a Stand when Tripping Pipe

Depending on the depth of the well, tripping pipe can take hours or even days, so it is not undertaken unless absolutely necessary. This is one reason why it is important to have a crew that knows how to properly utilize drilling fluid and take care of their equipment. One little mistake can set back a project considerably.

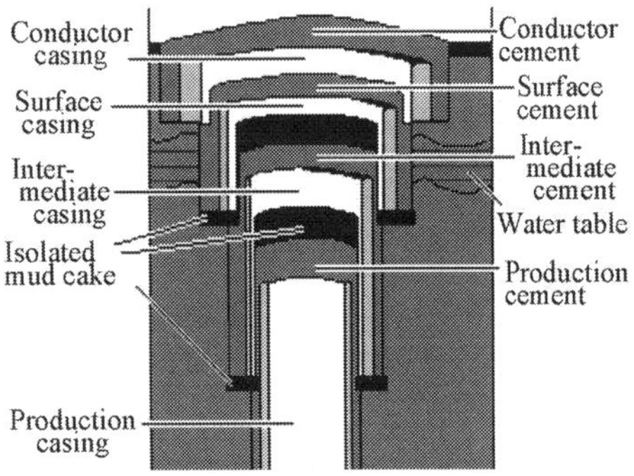

Figure 7 – 19
Layers of Casing and Cement

Casing

Surface and Conductor Casing

When it comes to the environment, protecting freshwater supplies is important. This is accomplished with a protective pipe layer called surface casing, and the depths at which it is required varies greatly as each state has different requirements. Once the well is just past the required depth for surface casing, the drill string is removed and the casing is lowered, welded and cemented in place. The pipe used for surface casing varies in diameter, is usually made of steel, and comes in various lengths that are connected together as they descend into the borehole. The first 20-50 feet of casing is called conductor casing. Conductor casing, also called conductor pipe or drive pipe, is cemented into place to prevent the well from caving in at the surface level.

Intermediate and Production Casing

In addition to surface and conductor casing, there is also intermediate and production casing, depicted in Figure 7 – 19. Intermediate casing is typically used if a borehole has multiple high-pressure zones, to prevent contamination or communication of fluids from one layer to another, and to stabilize the well. In addition, intermediate casing is commonly used with very deep and horizontal wells. Production casing is the last type of casing installed in a well and is unique in two ways: it spans the complete distance from the surface to the production zone, or target formation, and is the sole medium through which fluids are passed between the surface and the formation.

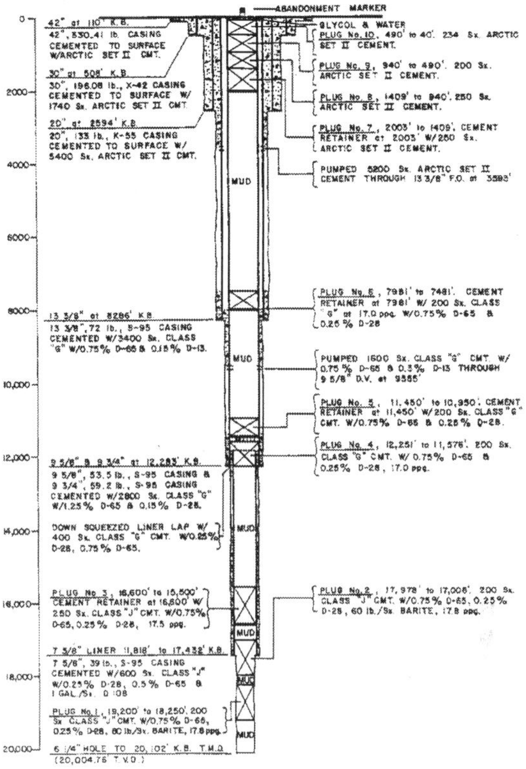

Figure 7 – 20
Detailed Casing Schematic
Source: U.S. Geological Survey (USGS)

Casing can be a somewhat confusing concept to those who are new to the industry. For example, one may look at Figure 7 – 19 and think that the production casing only "begins" further down along the borehole. This is not correct. Each stage of casing begins at the surface and fits completely within the previous stage. The casing layers are separated by cement each time they are installed and should never directly touch one another.

Because all of these pipe layers require ever-decreasing diameters in order to fit inside one another, they require varying diameter holes, the smaller the hole, the smaller the bit. This is another example of the vital importance of having a competent operator who knows how to order the correct bits for the project.

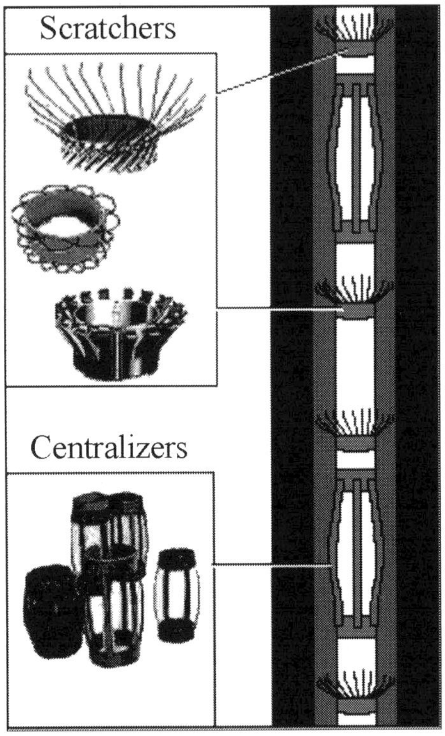

Figure 7 – 21
Scratchers and Centralizers

Figure 7 – 22
Connecting the Casing

Cementing the Casing

Correctly cementing the casing in place is one of the most important parts of the well, and proper steps should be taken to protect the environment when installing. Once the well is drilled slightly past the required depth for casing, the entire drill string is removed from the well. First, a guide shoe is attached to the bottom of the first casing segment. The guide shoe ends with a flapper, or dart, that prevents cement from flowing back inside the casing. A float collar is also installed above the guide shoe. Later, the float collar will be used to hold a rubber plug in place, which will be discussed shortly.

The casing is then lowered by the elevator one segment at a time. As this occurs, centralizers and scratchers are also installed around them. These devices are depicted in Figure 7 – 21. Centralizers keep the casing centered within the middle of the borehole around which cement can flow, thereby securing a straight well. It is important to have an even layer of cement between the casing and well walls to prevent possible operational and environmental issues in the future. Scratchers are designed to remove the drilling mud that is caked onto the sides of the well while casing is lowered. This allows the cement to form a stronger bond between casing and earth. During the scratching process, the mud-cake settles at the bottom of the hole. Planning for the isolation of this material by drilling beyond the desired target depth before installing the casing is vital.

This is the most environmentally safe way to case a well. It is only recently that strict requirements have been created that mandate the disclosure of all chemicals used in drilling mud. Isolating the scraped drilling mud at a non-hazardous depth, far from any freshwater aquifers, is the best scenario. As a result of these regulations, the fluids used during the conductor and surface casing stages are becoming more environmentally safe. The Environmental Protection Agency (EPA) imposes strict fines if any problems related to drilling mud should occur; and it for this reason that the operator is ultimately the one in charge of ordering and using the correct type of drilling fluids.

Figure 7 – 23
Cementing Head

Once the casing is in place, a well cementing company arrives on location and immediately begins to mix dry cement with water to a consistency that can easily be pumped through the inside of the casing and out and around the empty space between the casing and walls of the borehole. A cementing head is attached to the top of the last-installed casing segment, above the ground. A crew member releases a rubber plug called a *bottom plug* from the bottom of the cementing head. This device separates the cement from any remaining fluids within the casing. It is important to have pure cement to form the strongest chemical bond. When ready, the cement-water mixture, called slurry, is pumped under high pressure by the cement truck through the casing.

As the cement is injected, the bottom plug is pushed down. The bottom plug "seats," or stops, when it reaches the float collar. The bottom plug is designed to fit on top of, or even into, the float collar, but not to be pushed through. The cement crew knows when this plug is seated as there is a rapid build-up of pressure observed on the cement truck's gauges. This is caused by the continuation of cement being pumped.

The bottom plug is designed to break at a predetermined pressure. When the bottom plug gives way, the cement is pushed out the bottom of the casing and up the empty space between the casing and sides of the well, known as the annulus. Crew members know when the cement is moving toward the surface because there is a sudden drop in the pressure meter when this plug breaks. This drop is only temporary; the gauges quickly rebuild pressure as the cement continues to be forced up towards the surface.

After the bottom plug breaks, crew members release a second plug from the cementing head, called a *top plug*. The top plug is designed like the bottom plug but without any breakable passage. Once the top plug is released, the cement truck will typically pump a different fluid, such as water, into the casing. As more fluid is pushed into the casing, the top plug is pushed down towards the bottom plug. As this process occurs, cement remaining within the casing is forced out and pushed towards the surface. The cementing process is finished when the top plug meets the bottom plug, a contact that is referred to by workers as a *bump*. At this point, the equipment is disassembled, and the operator and drilling contractors get to rest while they Wait on Cement (WOC). WOC typically only takes a few hours depending on the depth. Figure 7 – 24 best describes this whole process.

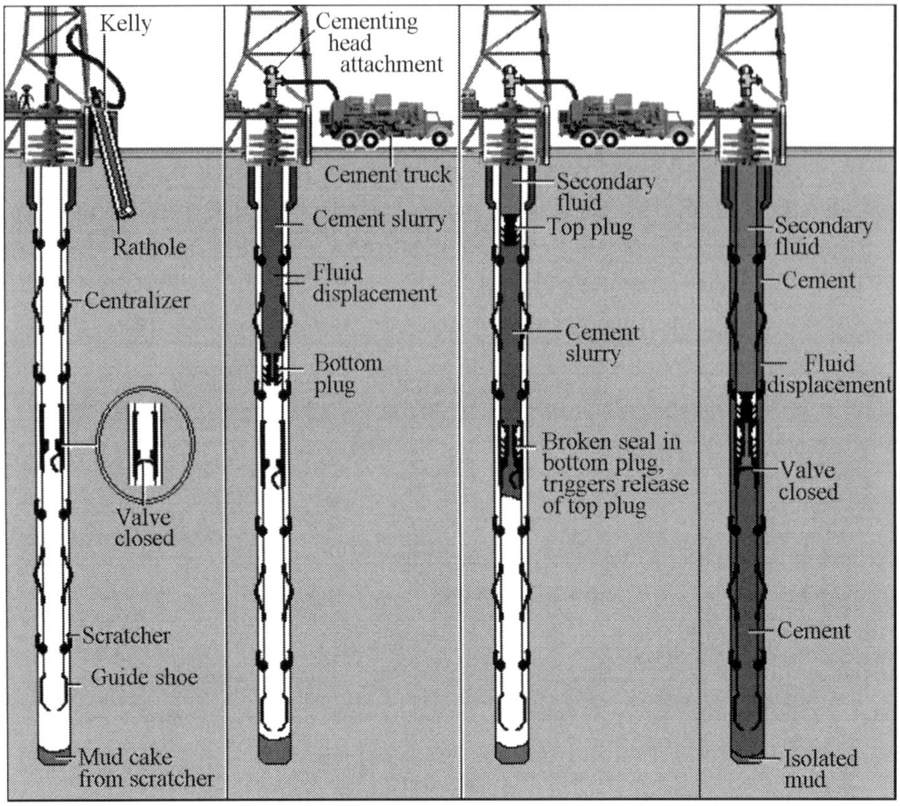

Figure 7 – 24
Cementing the Casing

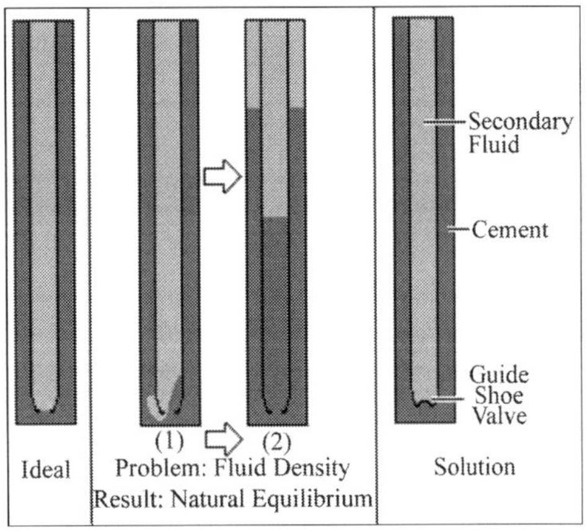

Figure 7 – 25
Ideal Fluid Displacement Using a Guide Shoe Valve

After WOC, a series of pressure tests are conducted to ensure the cement used to set casing is compliant with regulatory and industry standards. This is to make sure the casing can withstand the forces exerted by whatever is supposed to be kept out of the borehole.

Certain types of casing must pass these tests before drilling can continue or completion can occur. It is believed that these types of pressure tests showed signs of trouble on the Deepwater Horizon well in the Gulf of Mexico, which led to the failure of the well's blowout preventer, making the Gulf Oil Spill possible.

If there is any further drilling required once the casing has been installed, the tripping pipe process is reversed and the bit is lowered back to the bottom of the borehole. The bit drills through the bottom and top plugs and continues on its intended path.

Torque and Weight

The concepts of torque and weight are the applications of the lift and rotation forces described in the previous chapter. Torque is the force of rotation around an axis and can be thought of as twist, while weight is the downward force of gravity exerted on, in this case, heavy drill pipe and collars. A good example of the relationship between torque and weight as it applies to drilling is driving a screw into a piece of wood with a power screwdriver. If one does not have perfectly straight alignment between screwdriver, screw, and wood, or if too much or too little weight is applied to the task, the screw will rotate out of control and be driven in a corkscrewed fashion.

A straight hole test is performed, ideally every 500 feet, to determine if torque and weight have been properly applied. This is done by measuring any deviations from a perfect vertical line in the borehole. A tool called an inclinometer is lowered into the well where data is recorded and analyzed. If the deviation increases, the weight on the bit will be quickly decreased or pulled back. If the deviation is later observed to decrease, the weight on the bit can be adjusted again.

Misalignment is often caused by an object moving along the *path of least resistance*. As the drill bit grinds through the ground, it may encounter materials of unequal density, possibly leading to the bit being deflected in an undesirable direction. These deviations can increase substantially when drilling through materials like limestone. Many drillers will perform straight hole tests more often if they are encountering such situations. A hole with too much unwanted or unknown deviation can cause many problems in the future.

Common Drilling Problems

There are several problems that can occur during drilling if the crew is inexperienced or does not know what to look for. Most of the problems revolve around the mud circulating system of the rig. If the circulating system is faulty or is not monitored carefully, the cuttings may not be properly extracted; instead, they end up settling around the moving parts of the bit. After a period of time, this causes the bit to become stuck or locked up.

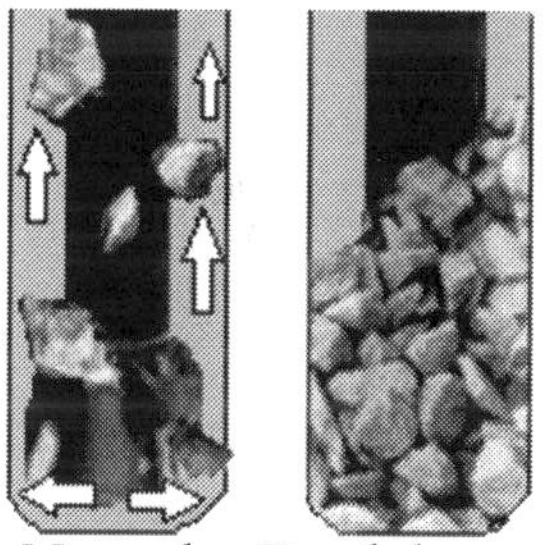

Figure 7 – 26
Drilling Breakdown due to Fluid System Failure

If improper fluid viscosity is present in the borehole for an extended period of time, the risk of the well collapsing in on itself increases. This can be a difficult problem if the drill string is still within the borehole. In some cases, the drill string may become completely stuck, impossible to remove from the ground. A common problem involving a corkscrewed well is an impediment that lodges the drilling string against the walls of the borehole, as seen in Figure 7 – 27. Impediments may also be seen toward the bend of a kick-off at the beginning of a horizontal drilling leg. These impediments can occur as the drilling fluid circulation dynamics are often uneven once a well deviates from being perfectly vertically straight.

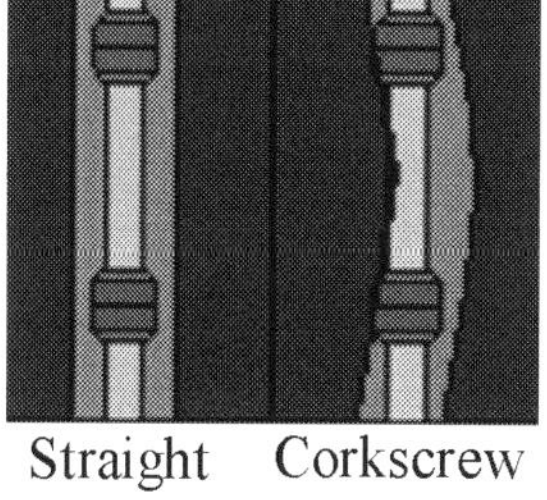

Figure 7 – 27
Corkscrew Effect

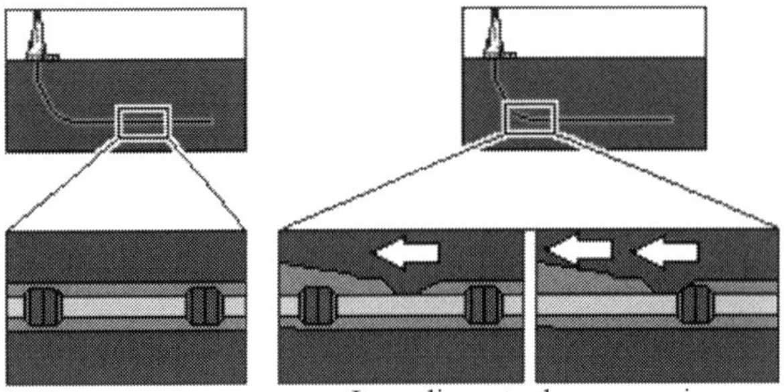

Impediment when removing horizontal drilling pipe

Figure 7 – 28
Stuck Pipe Scenarios for Horizontal Drilling

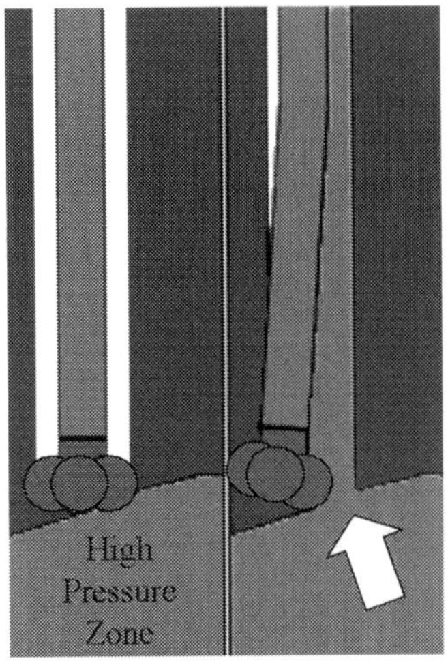

Figure 7 – 29
Downhole Blowout Causing Stuck Pipe

Sometimes a drill bit breaks through a formation and encounters an unexpectedly high level of subsurface pressure. The force of impact of such an encounter can lodge the drill string against the side of the borehole. Though these issues can be mitigated with intermediate casing, a scenario called a downhole blowout can occur if pressure is left unchecked. A human error-equivalent of this phenomenon can occur due to inappropriate monitoring or usage of the mud circulation system.

Two segments of pipe or collars can have their connection unwound by torsional stresses. The drill string beneath this point of disconnection is then twisted off and stuck at the bottom of the hole. This is a nearly disastrous scenario, but can sometimes be salvaged using a process called fishing. There are many different devices that can be used, but in the case of a twist-off it is common to use an overshot. Drill bits can also twist off from the collars and the effect and solution are the same.

In addition to a twist off, sometimes tools or other, smaller, undesired objects can also find their way into the borehole and lock up the drilling process. In this scenario the fishing process might call for a specialized object, such as a spear or a magnet, to be lowered into the borehole to retrieve the impediment. Failure to extract the item will cause the drill bit to attempt to chew into the metal object, which may eventually contribute to the failure of the bit.

Cracks can develop in pipes and collars and, when this happens, extreme caution is used when tripping for a cracked pipe as it can easily snap. To mitigate this risk, it is important that someone is in charge of inspecting all drill pipe and collars before drilling.

Another natural problem is entering a zone with a negative pressure that is of high porosity and permeability, causing a rapid loss of drilling fluids; this often happens when reentering old or depleted reservoirs.

Every moving part eventually breaks after long periods of usage. The bearings and teeth of the drill bit end up breaking or becoming too dull to be effective. This is a typical problem that should be considered by ordering bits *before* the drilling phase begins. Waiting for a bit to arrive because spare bits weren't ordered in advance is a terribly inefficient use of a drilling crew's time. This should be incorporated into the development plan by adding the cost of such contingency-based expenses.

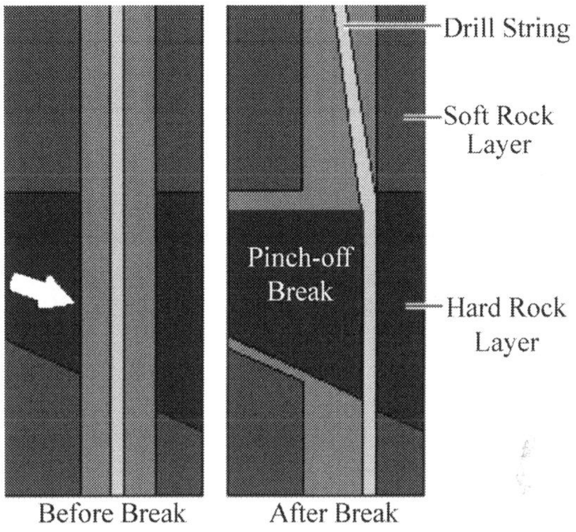

Figure 7 – 30
Pinch-off Break Causing Stuck Pipe

Pinch-off breaks can occur if one drills past an uneven formation. Over the course of drilling, vibrations from the drilling activity can dislodge a large piece of rock from the formation wall and into the pipe, locking the drill string between the walls of the well. Mud circulation is critical to prevent such a lock up.

Some problems are unavoidable, but many of them are caused by unsupervised human error. Problems, such as stuck pipe, can sometimes be solved by long periods of high pressure fluid circulation along with jarring the drill string. A device called a surface resonance vibrator may be used to try to dislodge a stuck pipe. If the problem persists beyond the financial budget for the well, the venture could collapse and issues could even contribute to environmental problems or fines. Once again, the vast majority of these problems may be averted by having experienced crews on these projects.

A crew that pays close attention to the rig's torque, weight and circulatory fluid pressure gauges is more likely to notice these issues before they develop into problems. Such gauges can serve not only as an indicator that something is wrong, but can aid in the diagnosis of what is wrong. Once the problem is defined, possible solutions can be created. If a situation occurs, it is critical to be as efficient as possible as mud can become more viscous and less effective as time passes.

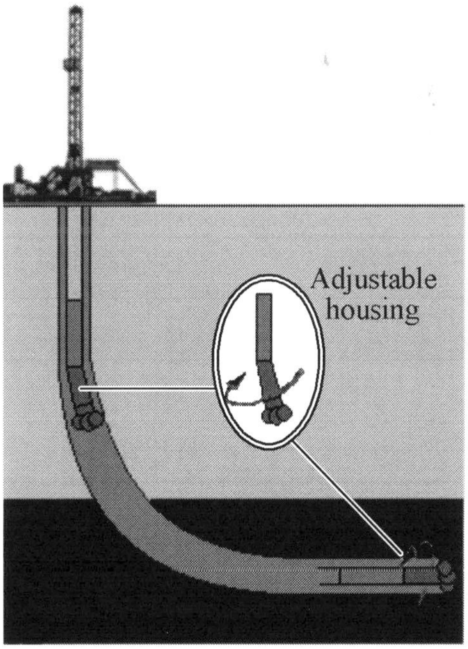

Figure 7 – 31
Downhole Drilling Motor with Adjustable Housing

Downhole Tools & Directional Drilling

One of the differences between vertical and directional drilling is the use of downhole equipment, commonly used for both onshore and offshore operations. Directional and horizontal drilling involves the use of a downhole motor to rotate the bit instead of using the turntable and kelly system located on the rig floor. The downhole motor uses the pressure of the passing drilling fluid to spin the bit, while the derrick-and-drawworks pulley system solely regulates the weight on the drill string.

The downhole motor has an adjustable housing which allows the motor to be bent at a variety of angles. These motors are typically attached to the drill string at a predetermined point above the formation to start the shift from vertical to directional drilling. Once the targeted deviation is achieved, the rotary turntable system is often used again. One advantage of downhole equipment is the ability to incorporate other tools, such as an inclinometer, location tracking devices, and a variety of sensors, just to name a few.

In horizontal drilling, the first step is often to drill a vertical hole to a depth below the target formation; this allows the exact determination of the formation's characteristics. For example, dip tests can be performed to better understand the formation shape. It is then determined whether it would be economical to go horizontal and, if so, at what height above the formation should the deviation occur. A steel wedge, called a whipstock, may be installed at the intended starting point of the deviation depth. This wedge aids the downhole motor when it is time to begin deviation.

Drilling Conclusions

Once the borehole reaches total depth (TD), the drilling phase is finished. An exploratory well drilled with little to no initial information may find a formation that is not porous or permeable enough to extract a sufficient quantity of hydrocarbons to pay for the next steps of a well's completion. It's even possible that the oil or gas is of such low quality that it is uneconomical to any nearby refinery.

As said numerous times throughout this book, although new exploration is high risk, it can be the most profitable activity in this industry. It is therefore extremely important to work with companies that fully understand all aspects of the drilling process.

Appendix A – Common Industry Units and Measures

Length

1 inch = 1/12 feet = 0.02540 meters = 2.540 centimeters
= 25.40 millimeters

1 foot = 12 inches = 0.000189 miles = 0.3048 meters

1 yard = 3 feet = 0.91440 meters

1 mile = 5,280 feet = 1,609.34 meters = 1.609344 kilometers

Weight

1 pound (lb) = 0.453592 kg

1 British/long ton = 2,240 pounds (lb) = 1,016.047 kg

1 U.S./short ton = 2,000 pounds (lb) = 907.1847 kg

1 metric ton = 2,204.623 pounds (lb) = 1,000 kg

1 pound-force = 0.450 kilogram-force = 4.450 newtons
= 0.4450 decanewtons

Pressure

1 PSI = 1 pound/square inch = 6,895.0 Pascal (Pa)
= 6.8950 kPa = 0.006895 MPa

1 PSI = 6.8948×10^{-2} bar = 6.8046×10^{-2} standard atmosphere (atm)
= 51.71493 Torr = 7.03069×10^{-2} technical atmosphere

Volume

1 barrel = 42 gallons = 5.615 cubic feet
= 0.1590 cubic meters = 159.0 liters

1 gallon = 0.13369 cubic feet = 3.79 liters = 0.003785 cubic meters

1 quart = 0.25 gallons = 0.033422 cubic feet
= 0.028317 cubic meters

1 cubic foot = 1728 cubic inches = 7.48 gallons
= 28.32 liters = 0.028317 cubic meters

1 fluid ounce = 29.570 milliters = 0.0295735 liters

1 pound per barrel = 2.8530 kilogram/cubic meter

Power and Energy

1 ft·lb = 1.3558 joules (J) = 13,558,000 Ergs (erg)
= 0.00128 British Thermal Units (Btu) = 0.3238 gram calories (cal)
= 0.0003238 kg cal or food calories

1 watt (W) = 44.2538 ft lb_f /min = 1 kg m^2/s^3
= $1x10^7$ erg/s = 106 megawatt (MW)

1 horsepower (mechanical) = 33,000 ft lb_f/min = 550 ft lb_f/s
=745.69988 W

Torque

1 ft·lb = 1.3558 newton meters

Viscosity

1 P = 0.1 Pa·s

1 cP = 1mPa·s = 0.001 Pa·s

Temperature

1 F° = 1 degree Fahrenheit = 26.666 Celsius (C°) = 299.8166 Kelvin (K)

Figures and Tables

All figures listed below are Copyright © Chris Termeer, all rights reserved, unless indicated as being in the Public Domain with an asterisk (*)

References

Ahmed, T. 2011. *Advanced Reservoir Management and Engineering* (2nd Ed.). Gulf Professional Publishing.

Altgelt, K. 1993. *Composition and Analysis of Heavy Petroleum Fractions.* CRC Press.

Australian Drilling, The. 1997. *Drilling: The Manual of Methods, Applications, and Management.* CRC Press.

Bannerjee, D. 2012. *Oil Sands, Heavy Oil and Bitumen: From Recovery to Refinery.* Pennwell Corp.

Beumont, E. 2000. *Exploring for Oil and Gas Traps.* American Association of Petroleum Geologists.

Brouwer, M. 2011. *Oil Refining and the Petroleum Industry.* Nova Science Publishers, Inc.

Chambers, A. 1999. *Natural Gas & Electric Power in Nontechnical Language.* Pennwell Books.

Chaudhry, A. 2004. *Oil Well Testing Handbook.* Gulf Professional Publishing.

Chavez, J. 2009. *Unconventional Oil: The Energy Transition Bridge.* VDM Verlag.

Cheresources Inc. 2010. "A Students Guide to Refining." Accessed May 13, 2012. http://www.cheresources.com/content/articles/energy/a-students-guide-to-refining.

Chhetri, A., et al. 2012. *Green Petroleum: How Oil and Gas Can Be Environmentally Sustainable.* Wiley-Scrivener.

Clark, G. 1988. *Industrial Marine Fuels Reference Book.* Butterworth-Heinemann.

Corbo, P., et al. 2011. *Hydrogen Fuel Cells for Road Vehicles.* Springer.

Dake, L. 1983. *Fundamentals of Reservoir Engineering.* Elsevier Science.

Devereux, S. 1997. *Practical Well Planning and Drilling Manual.* Pennwell Corp.

Donaldson, E., and Tiab, D. 2011. *Petrophysics: Theory and Practice of Measuring Reservoir Rock and Fluid Transport Properties* (3rd Ed.). Gulf Professional Publishing.

Drapcho C., et al. 2008. *Biofuels Engineering Process Technology.* McGraw-Hill Professional.

Elvers, Barbara. 2007. *Handbook of Fuels: Energy Sources for Transportation.* Wiley-VCH.

Enos, J. 2003. *Technical Progress and Profits: Process Improvements in Petroleum Refining.* Oxford University Press, USA.

Fennell, M. 2012. *Natural Gas and Other Hydrocarbon Technologies Explained Simply.* CreateSpace Independent Publishing Platform.

Freeburg, Russell. 1987. *Oil & War: How the Deadly Struggle for Fuel in WWII Meant Victory or Defeat.* William Morrow and Company, Inc.

Gadallah, M. 1994. *Reservoir Seismology*. Pennwell Books.

Green, D., and Perry, R. 2007. *Perry's Chemical Engineer's Handbook* (8th Ed.). McGraw-Hill Professional.

Gray, R. 1994. *Upgrading Petroleum Residues and Heavy Oil*. CRC Press.

Guibet, J. 1999. *Fuels & Engines, Vol. 1*. Editions Technip.

Hall, C., and Murphy, D. 2010. "Year in Review-EROI or energy return on (energy) invested," *Annals of the New York Academy of Sciences* 1185: 102-118. Accessed March 1, 2011. doi: 10.1111/j.1749-6632.2009.05282.x.

Harper, J. 1998. "Why the Drake Well?" *Pennsylvania Geology* 1:29, 2-4.

Holliday, Trevor W. 2010. *John D. Rockefeller 1839-1937: Industrial Pioneer and Man*. Kessinger Publishing, LLC.

Hubbert, M. 1956. "Nuclear Energy and the Fossil Fuels" *Shell Development Company* 95: 14-27.

Hyne, Norman. 2001. *Nontechnical Guide to Petroleum Geology, Exploration, Drilling and Production* (2nd Ed.). Pennwell Books.

Ingersoll, J. 1996. *Natural Gas Vehicles*. Fairmont Pr.

International Group, The. 2007. "Use of Microcrystalline Waxes in Candles." Accessed February 6, 2012. http://igiwax.com/uploads/articles/NCA Presentation-Microcrystalline Wax use in Candles.pdf.

Irwin, L. 2012. *Renewable Electricity Generation: Resources, Standards, Challenges*. Nova Science Publishers.

IUPAC. 1997. "Petroleum Coke," *Compendium of Chemical Terminology* (2nd Ed.). Compiled by McNaught, A., and Wilkinson, A. Blackwell Scientific Publications. Last modified August 19, 2012. doi:10.1351/goldbook.P04522.

Jahn, F., et al. 2008. *Hydrocarbon Exploration & Production, Volume 55* (2nd Ed.). Elsevier Science.

Johnson, F., and Rosillo-Calle, F. 2011. *Food versus Fuel: An Informed Introduction to Biofuels*. Zed Books.

Kaes, G. 2000. *Refinery Process Modeling*. Elliot and Fitzpatrick.

Kennedey, J. 1993. *Oil and Gas Pipeline Fundamentals*. Pennwell Books.

Kent, J. 2003. *Riegel's Handbook of Industrial Chemistry*. Springer.

Kidnay, A. et al. 2011. *Fundamentals of Natural Gas Processing* (2nd Ed.). CRC Press.

Kister, H. 1992. *Distillation Design*. (1st Ed.). McGraw-Hill Professional.

Knowles, R. 1980. *The Greatest Gamblers: The Epic of American Oil Exploration*. University of Oklahoma Press.

Leffler, W., and Miesner, T. 2006. *Oil & Gas Pipelines in Nontechnical Language*. Pennwell Books.

References

Lyons, W. 2009. *Working Guide to Drilling Equipment and Operations*. Gulf Professional Publishing.

Mansfield, C. 2000. *Base Oil Chemistry for Diesel Engine Lubricants*. Epsilon Enterprises

Marcel V. and Mitchell, J. 2006. *Oil Titans: National Oil Companies in the Middle East*. Brookings Institution Press.

Maugeri, L. 2006. *The Age of Oil: The Mythology, History, and Future of the World's Most Controversial Resource*. Praeger.

Mushrush, G., and Speight, J. 1995. *Petroleum Products: Instability and Incompatibility*. Taylor & Francis.

Mustafaoglu, O. 2009. *Characterization and Pyrolysis of Oil Shale Samples: An Alternative Energy Option*. LAP Lambert Academic Publishing.

Nash, K. 2010. *Shale Gas Development*. Nova Science Publishers Inc.

Occelli, M. 1996. *Hydrotreating Technology for Pollution Control*. CRC Press.

Parkash, S. 2009. *Petroleum Fuels Manufacturing Handbook: including Specialty Products and Sustainable Manufacturing Techniques*. McGraw-Hill Professional.

Picton-Turbervill, G. 2009. *Oil and Gas: A Practical Handbook*. Globe Law and Business.

Pillon, Lilianna. 2007. *Interfacial Properties of Petroleum Products*. CRC Press.

Pulkrabek, Williard W. 2003. *Engineering Fundamentals of the Internal Combustion Engine* (2nd Ed.). Prentice Hall.

Renpu, W. 2011. *Advanced Well Completion Engineering*. Gulf Professional Publishing.

Sadeghbeigi, R. 2000. *Fluid Catalytic Cracking Handbook* (2nd Ed.). Gulf Publishing.

Samuel, R. 2009. *Advanced Drilling Engineering: Principles and Designs*. Gulf Publishing Co.

Schmidt, P. 2010. *Fuel Oil Manual*. Industrial Press, Inc.

Seddon, D. 2010. *Petrochemical Economics: Technology Selection in a Carbon Constrained World*. Imperial College Press.

Simanzhenkov, V. 2003. *Crude Oil Chemistry*. Marcel Dekker.

Smith, M. 1970. *Aviation Fuels*. G.T. Foulis & Co, LTD.

Song, C. 2000. *Chemistry of Diesel Fuels*. CRC Press.

Speight, J. 2010. *The Refinery of the Future*. William Andrew.

Szilas, A. 1986. *Production and Transport of Oil and Gas: Gathering and Transportation*. Elsevier Science Publishing Company.

Venus Lun, Y., et al. 2012. *Oil Transport Management*. Springer.

Westbrook, M. 2001. *The Electric Car: Development and Future of Battery, Hybrid and Fuel-Cell Cars*. The Institution of Engineering and Technology.

Additional research material, primarily statistical in nature and attributable to no single body of work, was gathered from the following agencies and organizations:

American Association of Petroleum Geologists

American Petroleum Institute

ASTM International

BP

Bureau of Land Management

Chevron Corporation

Department of Energy

Energy Information Administration

Environmental Protection Agency

Federal Highway Administration

International Energy Agency

International Standards Organization

Library of Congress

National Aeronautics and Space Administration

National Oceanic and Atmospheric Administration

New York Mercantile Exchange

Occupational Safety and Health Administration

Organization of the Petroleum Exporting Countries

Platts

SAE International

Society of Petroleum Engineers

United States Air Force

United States Geological Survey

United States Patent Office

Glossary

A

absorption: A chemical or physical reaction whereby a substance is "consumed" and distributed throughout another substance (solid, liquid or gas).

acidizing: A type of well stimulation/completion technique in which acids are used to chemically break down soluble material of a subsurface rock matrix, increasing formation permeability and porosity. For example, hydrochloric acid (HCl) can break down limestone.

additional treatment: Third major stage of oil drilling which includes the processes of desulfurization, blending, etc

additive: A substance, usually a liquid, added to a finished product to alter specific chemical properties. For example, ethanol is said to be an "octane-improving additive" to gasoline.

adsorption: Similar to absorption; however, the consumed substance is only distributed across its surface only, not throughout its interior.

AFE: *see Authorization for Expenditure*

alkaline surfactant polymer (ASP) flooding: A secondary recovery method in which various alkaline chemical compounds, soaps, and gases are injected into the reservoir in an attempt to "flush out" any remaining hydrocarbons.

alkylate: The output of a refinery's alkylation unit. Alkylate is independently traded as a high-octane gasoline blendstock.

alkylation: An enhanced refining process that combines smaller hydrocarbon molecules, usually in the presence of a catalyst, to create larger hydrocarbons more suitable for blending into gasoline.

alternative energy: Energy sources other than fossil fuels (coal, crude oil, natural gas) used for electricity generation, transportation, or other uses. Alternative energy generally includes biofuels, renewable energy and nuclear power.

American Petroleum Institute (API): A U.S. trade organization that gathers statistical information about U.S. production and attempts to self-regulate the industry.

American Society for Testing and Materials: *see ASTM International*

amine washing: The removal of an H_2S stream as the last phase of hydroprocessing or hydrodesulfurization

annulus: The empty space between the wellbore and drill string, when drilling, or between production tubing and casing when a well is producing.

Anti-Knock Index (AKI): The average of a fuel's Research Octane and Motor Octane Numbers. This number is prominently displayed on a yellow sticker at all U.S. gas stations.

API: *see American Petroleum Institute*

API gravity: An arbitrary but widely used specific gravity scale. API gravity = (141.5/SG at 60°F) - 131.5, where SG is the specific gravity of the fluid.

appraisal well: An oil exploration technique involving the use of vertical (or directional) boreholes to determine the outside edges of a subsurface reservoir.

aquifer: A subsurface rock formation containing liquid water within its pores in a manner somewhat similar to how crude oil is found in an oil formation.

Arab Spring: Common name for a series of protests that swept the Middle East and North Africa throughout 2011. These events are widely believed to be attributed to high global food prices, including corn.

area drilling superintendent: Essentially a "district manager" for a drilling company, this individual supervises operations for multiple drilling rigs, generally within a limited geographic area.

aromatic: A type of hydrocarbon molecule made of a ring structure.

As-Is Agreement: Arrangement between the so-called Major oil companies, signed in 1928, in an attempt to control global oil prices.

asphalt: Commonly used interchangeably with the terms bitumen and binder, asphalt is actually the mixture of bitumen with aggregates, which are made of crushed rock, stone, and sand.

asphalt concrete: *see asphalt*

asphaltene: Extremely large, dense, complex hydrocarbon molecules. Asphaltenes may contain over 70 carbon atoms per molecule and are commonly found in higher density oil.

assay: A chemical analysis of the properties and composition of a substance. A crude oil assay will determine its density, viscosity, contaminant ratios, and other properties.

associated gas: Any natural gas freely produced from a crude oil well or dissolved within the liquid mixture.

ASTM International: An international organization that publishes engineering and technical standards, including test methods and best practices.

atmospheric distillation: Generally the first step of crude oil refining in which crude oil is heated and its various hydrocarbon molecules are vaporized and later collected according to their approximate boiling range.

atom: The most basic unit of matter under classical chemistry. Atoms consist of varying combinations of neutrons, protons and electrons.

Authorization For Expenditure (AFE): A comprehensive budget for the drilling of a well. A properly drafted AFE should include costs for labor, the drilling rig's rental rate, pipe, bits, cement, mud and many other costs, as well as a contingency for unknown and likely unavoidable expenses.

autoignition: An automotive malfunction characterized by fuel being ignited by the ambient heat within the engine, rather than by the spark plug. Commonly confused with knocking.

avgas: *see aviation gasoline*

aviation gasoline: A gasoline-like product used by general aviation aircraft (small propeller-driven airplanes and most helicopters). The distinguishing feature of avgas, compared to motor gasoline, is its lead content.

B

bail: A downhole device used to extract tools or either debris from the borehole. They are critical to cable-tool drilling rigs as they have no fluid circulation system to remove cuttings from the borehole.

ball and seat valve: A downhole assembly used by some artificial lift systems to control fluid flow between the wellbore and production tubing.

barrel: A standard unit of measuring liquid crude oil. 1 standard barrel = 42 U.S. gallons.

base gas: Minimum volume of natural gas needed to maintain an underground storage system. Also called cushion gas.

base oils: Refinery byproduct used as an input in the creation of a wide variety of finished lubricant formulations, greases and other consumer products.

basic sediment and water (BS&W): The amount of water and various solid particulates not removed at the oil production site, typically expressed as a percentage of the sample's weight.

basic separation: The first stage of oil refining, typically involving atmospheric and vacuum distillation.

benchmark pricing: The use of a reference blend of crude oil for pricing other, less common blends. Benchmark blends include WTI and Brent North Sea.

Big Oil: A commonly used term used to describe either the Majors or, more broadly, the oil and gas industry as a whole.

bioasphalt: A type of asphalt derived from vegetable oil made from a variety of food products

biodiesel: An alternative fuel source made by the transesterification of vegetable oils or animal fats.

biomass power: Utility-scale electricity generation from the combustion of wood, solid waste, and various types of agricultural waste, even garbage.

bitumen: A high density crude oil byproduct commonly used as a binder to make asphalt. The term is also used to refer to Canadian oil sands heavy crude oil (natural bitumen).

bitumen blowing: A method of bitumen production that involves forcing hot air through straight run residue. This improves the physical properties of the product in a way that makes it better suited to making asphalt.

black gold: A term used to associate the early years of the crude oil industry to the great California "gold rush" of the 1840s.

black oil: A general term for both heavy distillates and for raw crude oil itself as they are less volatile and more dangerous if spilled than so-called white oils, which are more volatile and would naturally evaporate into the atmosphere if spilled.

black oil pipeline: A pipeline that transfers raw, unrefined crude oil as opposed to one that transports finished petroleum products,

blending: One of the final stages of the refining process that involves combining various intermediate products to create finished products.

blends (of crude oil): A collection of crude oil from a variety of individual wellheads and fields with similar chemical characteristics. Also called grades, individual blends of crude oil can be sold on the global market in such a way that the buyer generally knows what he's getting.

blendstock: Generally, any input used to manufacture a finished product. Many gasoline blendstocks are independently traded on the global market.

blowout: An undesirable explosion of crude oil to the surface. Alternatively, a downhole blowout involves the flow of reservoir fluids into a different subsurface structure or formation.

blowout preventer (BOP): A series of valves installed during the drilling process that, if closed, will prevent an uncontrollable explosion of fluids to the surface.

BOB: *see Bottom of Barrel index*

BOP: *see blowout preventer*

borehole: A vertical or horizontal hole artificially made in the ground, usually with the intention of extracting resources, such as water or hydrocarbons.

bottom dead center (BDC): The point in a pumpjack's stroke when the horsehead is nearest the ground.

Bottom of Barrel index (BOB): A method for classifying individual refineries, or entire oil companies, according to the degree to which they can process higher density crude oil blends.

box: The "female" end of a segment of drill pipe's tool joint.

Brazil, Russia, India and China (BRIC): The BRIC bloc is often discussed in economics as a high-growth region with positive trade balances, growing populations and, essentially, good investment opportunities.

Brent North Sea blend: The European standard benchmark crude oil blend frequently cited by market commentators and analysts. Brent oil comes from four distinct fields, is traded in London, and physical delivery occurs in Sullom Voe, Scotland.

BRIC: *see Brazil, Russia, India and China*

BS&W: *see basic sediment and water*

BTX (benzene, toulene, xylene): A mixture of high density aromatic hydrocarbons. BTX molecules are used as feedstock in the creation of many petrochemicals, including nylon, styrene, acetone and polyurethane.

bubble cap: Microscopic holes in a refinery's distillation unit's horizontal trays through which only vapors, not liquids, can pass.

bubble point: A level of reservoir pressure (measured in psi), above which, natural gas dissolves within the liquid crude oil.

bullwheel: Component of a cable-tool drilling rig's pulley system.

bunker fuel: Marine industry jargon for any type of fuel used by ocean-faring vessels. These ships commonly utilize liquefied natural gas (LNG) or various blends of diesel and residual fuel oil.

bunkering: *see bunker fuel*

butane (C_4H_{10}) (C4): One of the heaviest hydrocarbon molecules typically associated with natural gas. Butane is used both as a fuel source and petrochemical feedstock.

byproduct : An output of a chemical or physical process. Many byproducts are then used as inputs into other processes before becoming an end-consumer product.

C

cable-tool drilling: An early form of oil drilling in which a weighted chisel is repeatedly raised and dropped down a single vertical hole.

cake: Drilling fluid attached to the side of a borehole. Cake is useful during drilling because it helps prevent borehole collapse, but must be removed before the well comes online.

calcined coke: A form of petroleum coke more highly refined than the coker's raw output. Calcined coke has lower water and volatile hydrocarbon content than green coke.

California Air Resource Board (CARB): State agency tasked with regulating motor fuel pollution standards within its borders.

cap rock: Layer of impermeable rock situated above a hydrocarbon-bearing formation. Cap rocks trap the hydrocarbons in place, else they may drift up to the surface as a seep.

carbon black: Produced from crude oil refining, carbon black consists of almost pure carbon, and is used in the manufacturing of rubbers and inks.

carburetor: A type of fuel injection system used in gasoline engines, generally before the 1980s. Today, carburetors can still be found in smaller engines, such as dirt bikes and jet skies.

cased-hole completion: A type of completion in which a length of production casing is lowered to the depth of the intended oil-bearing formation. This casing must be perforated before reservoir fluids can enter the borehole.

casing: Any of various types of hollow steel pipe lowered into and cemented against the sides of a borehole. Generally, casing is intended both to prevent well collapse and to protect different formations from contaminating each other.

catalytic reforming: *see reforming*

cathead: Auxiliary winches powered by the rotational motion of the drawworks on a drilling rig. Among its more important functions is tightening and loosening pipe connections throughout the drilling phase.

CBOB: *see conventional blendstock for oxygenate blending*

cellar: A shallow underground concrete- or wood-lined work area located around - and including - the wellhead, directly underneath the rig floor.

centralizers: Mechanical devices wrapped around segments of casing to ensure they are properly centered within the borehole.

cetane index: Alternative method of calculating diesel fuel's cetane number according to a sample's density and distillation temperature profile.

cetane number: A quality rating scale for diesel fuel similar to octane rating gasoline. In this case, the measured effect is the fuel's tendency to self-ignite due to compression.

chromatography: Refers to a scientific technique that measures the relative proportion of various substances within a mixture, such as natural gas.

Claus Process: A method of removing pure elemental sulfur from a natural gas stream.

Clean Air Act: A major body U.S. environmental legislation responsible for regulating air quality, vehicle emissions standards, ozone protection and many other things.

clean pipeline: A pipeline that transfers finished petroleum products, as opposed to one that transports raw, unrefined crude oil.

CNG: *see compressed natural gas*

coal: A dense, solid hydrocarbon rock structure created from geologic processes similar to those which create crude oil. Commonly burned as a source of electricity.

coalbed methane: Gaseous methane adsorbed *along* the inner surface of coal's pores. Can be extracted from subsurface coal seams by relatively conventional gas drilling techniques.

coking: An enhanced refining process that applies increased temperature and pressure to straight run residue in order to split large hydrocarbons into smaller ones. Petroleum coke is a byproduct.

cold mix asphalt: A broad term for several asphalt grades with lower mixing and operating temperatures than traditional hot mix asphalt

collars: Heavily weighted lengths of hollow steel pipe intended to apply weight on the bit.

Colony Collapse Disorder: A term used to describe the unknown and rapid disappearance of honeybee colonies in the United States.

combustion: A type of chemical reaction requiring an oxidizing agent and a fuel. Combustion, also called burning, typically produces heat and/or light, and facilitates modern transportation.

commodity: Generally used to refer to a class of broadly sold goods that are more or less interchangeable throughout the market. Examples include coffee, soybeans, beef, silver and crude oil.

compaction drive: A relatively uncommon method of crude oil extraction that relies on a reduction of the reservoir's total volume, which is caused by collapsing reservoir rock.

company man: A representative of the operating company present during a well's drilling phase. The company man is often responsible for coordinating the various contractors and specialists that will be needed.

completion phase: That stage of bringing a new well online during which time production equipment is first installed and one or more well stimulation tasks are performed.

compound: A combination of two or more elements held together by a chemical bond.

compressed natural gas (CNG): A refined product consisting primarily of methane, natural gas used in this manner is highly compressed to reduce its volume and is often used as a transportation fuel.

concession agreement: A type of legal agreement, typically between landowners, or public governments, and oil companies. In it, the right to extract oil and gas is exchanged for a certain percentage of the proceeds of that activity.

condensate: *see natural gas liquids*

condenser: A device used within an atmospheric distillation unit to capture, cool off, and eventually reinsert any vapors which did not condense on a lower tray level within the unit.

conductor casing: Highest-diameter segment of casing, typically installed to protect the integrity of the borehole at the surface.

congeal point: The highest temperature at which a melted solid re-solidifies, an important chemical characteristic of petroleum waxes and other similar products.

contingency: A budgeted line item in an Authorization for Expenditure that would cover a wide variety of unforeseen scenarios during the drilling and completion of a well.

conventional blendstock for oxygenate blending (CBOB): Quoted on the financial markets as CBOB, the term refers to conventional gasoline bought and sold between refiners and wholesalers prior to the addition of ethanol.

conventional gasoline (CG): The most basic type of finished motor gasoline subject to the least stringent environmental standards under federal regulations.

conventional oil and gas: Oil and natural gas found within the pores of subsurface rock formations, extracted by drilling a borehole to the reservoir and requiring relatively little additional effort to do so.

Coordinated Emergency Response Measures: The oil importing world's combined strategic petroleum reserves. This term refers to the collection of IEA member nations' required stockpiles of crude oil.

core sampling: The extraction or removal of a piece of formation rock (sample) from a potential reservoir for further study.

corkscrew effect: The tendency of a drill bit to deviate from a straight (vertical or horizontal) line, usually according to the path of least resistance.

crack spread: Essentially a measure of a refinery's gross product margin calculated by subtracting the finished product selling price from the cost of the crude oil input.

cracking: Any of a number of enhanced refining processes in which large hydrocarbons are chemically broken into two or more smaller molecules better suited for blending into profitable transportation fuels.

crown block: The permanent sheaves located at the top of a drilling derrick essential to the rig's lift system,

crude oil: *see oil*

cryogenic expansion: A method of separating the individual components of natural gas by lowering the temperature of the mixture and isolating methane as the heavier molecules condense into liquid.

cutback bitumen: Bitumen diluted with petroleum fractions, such as kerosene or naphtha, in order to facilitate transportation and handling.

cuttings: Bits of rock separated from the earth by the grinding or scraping motions of the drill bit.

D

deasphalting: The process of removing complex asphaltene molecules from straight run residue, generally with the aid of a solvent.

Deepwater Horizon: An offshore drilling rig involved in the 2010 BP oil spill in the Gulf of Mexico.

dehydrogenation: An enhanced refining technique in which some hydrogen atoms are removed from a hydrocarbon molecule, converting saturated alkanes to unsaturated alkenes in the process.

demand destruction: An economic phenomenon whereby the aggregate demand for a product is reduced over the long term in the face of sustained high prices.

denature: To render undrinkable. This is usually done to fuel ethanol by adding toxic compounds, such as methanol, acetone, or petroleum liquids like naphtha.

density: The mass per unit volume of a substance. Density is one of the most important physical properties of crude oil and many of its finished products.

deoiled wax: Byproduct of the solvent dewaxing process used to manufacture finished wax products, including candles.

depletion: The process of an oil or gas reservoir having its contents extracted over time.

depth of oil water contact: The vertical depth within a reservoir at which separated oil and water interact with each other.

derrick: A tall, typically four-legged steel structure that supports the weight of the drill pipe, collars and bit when drilling a new borehole.

derrickman: Worker on a drilling rig who, during the tripping pipe process, specializes in lining up individual stands within the fingers of the monkeyboard at the top of the rig.

desalter: Component of a refinery's basic separation system that uses water and heat to dissolve salt from a crude oil stream into the water prior to atmospheric distillation.

developed reserves: A sub-classification for any quantity of proven, probable, and possible reserves that do not require significant amounts of capital to extract.

dewatering: In the upstream industry, this term refers to the removal of water dissolved in produced crude oil before selling it to a refinery. This is accomplished by two- or three-phase separators in located in the field.

dewaxed oil: Byproduct of the solvent dewaxing process used to manufacture base oil, lubricant and grease feedstocks.

dewaxing: Enhancement refining process that uses a solvent to isolate paraffin wax from straight run residue.

diesel engine: A type of internal combustion engine wherein fuel is self-ignited by the ambient heat and temperature conditions within the combustion chamber.

diesel fuel: Developed by Rudolph Diesel in 1893 to fuel his new engine design. Diesel fuel is a middle distillate of higher density than gasoline.

dip: Refers to a subsurface rock formation being angled relative to true horizontal.

dip water flooding: A water flood strategy that takes advantage of a slanted oil reservoir, and the earth's gravity, by placing the producer well "down dip" of the water injector well.

directional drilling: Any of a number of non-vertical well drilling strategies, including horizontal. Directional wells are useful in reaching reservoirs not directly accessible from the surface above, such as underneath a lake or other property with surface access issues.

disposal well: A well drilled, or an existing well converted, for the purpose of safely injecting fluids under the ground, such as frac fluids.

distillation: A chemical process used to separate components of a liquid mixture according to differences their boiling points.

distillation temperature profile (DTP): A type of test that describes the hydrocarbon content of a sample by reporting the temperatures at which 10, 50 and 90% of the sample evaporates. DTP is closely correlated to density for both crude oil and its finished products.

doghouse: Modular structure erected on the rig floor serving as a multi-purpose room to the drilling crew.

downcomer: In refining, a flow of liquids within an atmospheric distillation unit to a lower tray level according to each of the hydrocarbons' molecular density.

downstream sector: The processes, facilities, personnel and infrastructure dedicated to the creation of refined crude oil and natural gas products and delivering them to consumers.

draw: The difference in pressure between the reservoir and the bottom of the wellbore. Also, as a verb, refers to the actual lateral movement of hydrocarbons across the reservoir itself.

drawworks: An auxiliary motor on a drilling rig that controls its lift system by reeling in and letting out its drilling line.

drill bit: The component of a drilling rig that actually cuts into rock as it gets lowered deeper into a wellbore, facilitated by the weight of the pipe above and the rotational motion of a rotary turntable, top drive or some other mechanism.

drill pipe: Segments of steel, hollow cylinders with threaded fittings on the ends so they can be connected together as the bit descends further away from the rig and into the ground.

drill pits: Shallow holes or above ground tanks located at the drilling site used to collect cuttings and other debris.

drill stem test (DST): A type of reservoir analysis technique in which fluids from the targeted reservoir are allowed to flow to the surface on its own, facilitated by a system of surface valves and downhole plugs.

drill string: An umbrella term for the combined length of drilling collars, pipe and any downhole equipment, excepting the bit itself.

drilling: The process of boring a hole into the ground, usually with the intention of extracting resources such as oil, natural gas, or water.

drilling fluid (mud): A mixture of water, dirt and additives injected through the hollow drill string, out through the bit, and into the borehole. Drilling fluid is the "blood" of the rig's circulation system.

drilling foreman: The senior position on the rig floor under the toolpusher. Among other roles, the drilling foreman primarily operates the rig's mechanical equipment, including the drawworks.

drilling line: Braided steel cables connecting the various parts of the drilling rig's pulley system.

drilling pad: A temporary or permanent level surface that serves as a work area for a rig's personnel and equipment.

drilling phase: That stage in which a hole is created to make a well.

drilling rig: A collective term for a derrick and all accompanying systems, motors and equipment necessary to bore a hole into the ground.

drip gas: *see condensate*

drive pipe: *see conductor casing*

dry gas: This term refers both to natural gas which is almost entirely consisting of methane, and to natural gas which has no water content.

dry hole: A drilled well which is not able to produce an economically viable quantity of hydrocarbons. Ideally, a dry hole is identified early so that it can be plugged and abandoned prior to completion.

DST: *see drill stem test*

DTP: *see distillation temperature profile*

E

E&P: *see exploration and production company*

E85: *see gasohol*

easy oil: Refers to shallow, onshore oil reservoirs also commonly referred to as conventional oil.

EIA: *see Energy Information Administration*

element: A pure substance consisting of one or more atoms with the same number of protons and neutrons.

elevators: The component of a drilling rig that mechanically clamps onto the segment of drill pipe currently entering the hole.

emulsified bitumen: A type of cold mix asphalt that uses water mixed with bitumen with the aid of a surfactant. This results in improved handling and operating properties.

emulsified fuel: Heavy oil products can be mixed with water and a surfactant to create a relatively clean burning transportation fuel.

emulsion: A compound of two or more insoluble liquids 'held together' by a surfactant.

energy density: Essentially, the amount of work that can be performed per unit volume of a substance. It is similar to concept of specific energy and is used to provide an "apples-to-apples" understanding of different fuels and energy sources.

Energy Information Administration (EIA): Sub-agency of the U.S. Department of Energy tasked with, among other things, collecting statistical data about the country's energy trends, including production, consumption, imports, exports and technology.

energy return on investment (EROI): The foundation of energy economics, it is a simple ratio of the amount of energy used to create additional energy.

enhancement: The second stage of oil refining during which time the various distillate fractions are further processed. The ultimate goal of enhancement is to maximize the output of profitable finished goods like gasoline, jet fuel, diesel fuel and home heating oil at the expense of undesirable products like residual fuel.

Environmental Protection Agency (EPA): Federal agency tasked with regulating and enforcing environmental laws. Areas of authority include water, air, motor vehicle pollution and insecticide and pesticide use, to name just a few.

EPA: *see Environmental Protection Agency*

EROI: *see energy return on investment*

ethane (C_2H_6) (C2): A gaseous paraffinic hydrocarbon making up the second largest component of raw natural gas. Common petrochemical feedstock.

ethanol fuel: A colorless, flammable liquid produced from fermented sugars used as a transportation fuel in lieu of (or blended with) finished motor gasoline or other fossil fuels. Unlike drinking alcohol, fuel alcohol must be denatured by its manufacturer before its sale.

evaporation: The process of a liquid becoming a gas, either due to changes in temperature, pressure or some combination of both.

excavation: The clearing of land around a potential or pending drilling site. In some cases, the building of roads or other infrastructure is included in this activity.

exploration: An umbrella term that encompasses a wide variety of activities necessary to bring a new oil or gas well online. These include researching suitable locations, conducting seismic or other studies, even the drilling of an exploratory well and analysis of its resulting data.

exploration and production company (E&P): A company that focuses its activities and efforts solely on bringing additional oil to market by finding new reservoirs and extracting their hydrocarbons.

F

FAA: *see Federal Aviation Administration*

FCC: *see fluid catalytic cracking*

Federal Aviation Administration (FAA): A national agency tasked with regulating flight safety, pilots, traffic control, and similar functions and tasks.

Federal Energy Regulatory Commission (FERC): An independent agency under the Department of Energy tasked with regulating interstate energy commerce, including pipelines, electricity, export terminals, markets and safety.

feedstock: Any input into a refining process. Note that some crude oil finished products are, in turn, feedstocks to others outside the oil industry. For example, petroleum wax is a feedstock for candle manufacturers.

FERC: *see Federal Energy Regulatory Commission*

fertilizer: Generally refers to any product that delivers nutrients to the soil and farmland with the intention of increasing crop yields. Ammonia, a petrochemical, is one of the primary ingredients in many common fertilizers.

finished product: *see petroleum-based products*

Fischer-Tropsch Process (FT): Formerly a patented chemical process, this term now refers to a series of chemical reactions that artificially create paraffinic molecules (also called alkanes) suitable as feedstock for manufacturing a variety of transportation fuels, such as jet fuel, diesel, or gasoline.

fishing: This term refers to any process or tool used to retrieve lost or damaged equipment from the bottom of the borehole.

fixed drill bit: A type of rotary drilling bit that has no moving parts. Its cutting action is said to be more of a "grinding" force.

flaring: Trace amounts of natural gas from a geographically isolated oil well is difficult to bring to market. Traditionally, this gas was separated from crude oil and burned directly at the field. This practice is now illegal in the U.S except in temporary circumstances.

flash point: The temperature above which a substance can be ignited by ambient heat, that is, without requiring a direct ignition source.

float collar: A downhole device used during the cementing of surface casing to prevent fluids from flowing back up through the borehole.

fluid catalytic cracking (FCC): A type of cracking process that uses a solid catalyst and extremely high temperatures to break large hydrocarbons into smaller ones.

flush production: The stage of an oil or gas well's production cycle in which hydrocarbons are able to reach the surface under the reservoir's own pressure, without the need for artificial lifting equipment.

food versus fuel debate: An umbrella term for the issues related to diverting crops, such as corn, soybean and sugar cane, for the production of biofuels, such as ethanol and biodiesel. For example, corn sold to make ethanol reduces the supply of corn available in the food market, ultimately making corn more expensive to food consumers.

formation: A distinct body of rock strata commonly named after its rock type or location, or both. Formations are also sometimes referred to as zones.

formation volume factor: The volume that a standard barrel of oil occupies under reservoir conditions. This volume differs from surface conditions due to the increased subsurface pressure.

fossil fuel: Energy derived from decomposed plant and animal remains, compressed into increasingly energy-dense forms over time. Included in this category are coal, natural gas and liquid crude oil.

frac: *see hydraulic fracturing*

fracking: *see hydraulic fracturing*

fraction (of crude oil): *see intermediate product*

fractionators: Essentially, a re-distillation unit; a fractionator is used at the end of the coking process to separate the resulting hydrocarbons into similar intermediate products as are found during basic separation.

fuel ethanol: *see ethanol fuel*

free flow: *see flush production*

fuel injector: A component of an internal combustion engine that controls the flow of vaporized fuel into the combustion chamber with a series of valves.

fuel oil: An umbrella term for a wide variety of petroleum fuels derived from the middle and heavy ends of the barrel. Included under this classification are diesel fuel, kerosene, home heating oil and residual fuel oil

fungible: A product is said to be fungible when it is more or less interchangeable with a similar product from another manufacturer. In the petroleum industry, this is largely accomplished with the use of standards.

futures contract: An exchange-traded contract for the delivery of a commodity at a specified date and location at an agreed-upon price.

futures exchange: A marketplace upon which futures contracts are bought and sold. Such markets provide price stability and transparency, are relatively high in trade volume, and reduce the volatility of the market as a whole.

G

gas cap drive: Where fluids from the reservoir are expelled to the surface by the expanding forces of free and dissolved gases within the reservoir itself.

gas coning: Occurs when free gas is pulled to the wellbore more quickly than liquid oil, displacing the oil along the permeable channels, resulting in a distorted gas-oil extraction ratio.

gas fractionation: A refining process in which the components of natural gas are separated from a mixture into their individual, pure forms.

gas injection: Where various types of gases are injected into a formation as a way to increase reservoir pressure. Typically used in secondary recovery.

gas lift pump: A mechanism used to extract hydrocarbons from a reservoir. With it, compressed gas is used to alter the pressure and overall fluid dynamics within the wellbore.

gas separator: Equipment used at the extraction field to separate gas and liquids produced as a mixture from a well.

gasohol: A gasoline blend that contains alcohol, such as ethanol. Its grades are commonly referred to according to their alcohol content ratio. For example, E85 refers to a blend containing 85% ethanol, 15% finished motor gasoline.

gasoline: A relatively volatile refined product of crude oil that contains a wide mixture of varying hydrocarbons and additives. Commonly used as a fuel for an internal combustion engine.

GDP: *see gross domestic product*

Generally Accepted Accounting Principles (GAAP): A set of standards, guidelines, rules and procedures pertaining to accounting practices set by the Federal Accounting Standards Board (FASB).

generator (of electricity): A device that produces electricity by converting mechanical energy into electrical energy.

geological column: A special presentation of drill cutting samples arranged according to depth taken.

geothermal power: Utility-scale electricity generation created from fluids extracted from a underground reservoir. The mechanical energy from the hot water or steam is then used to power a generator.

Ghawar Field: The largest conventional oil field in the world. Located in Eastern Saudi Arabia, this "elephant field," as such large fields are called, allegedly produced 5 million barrels of oil per day.

glycol dehydration: A procedure that uses glycol as an absorbent to remove trace amounts of water from natural gas.

grades (of crude oil): *see blends*

gravel packing: Where a screen mesh and gravel of a particular size is attached to the bottom of the well to prevent the passage of sand from the formation to the wellbore.

gravimetric survey: A hydrocarbon exploration technique that measures density variations in subsurface rock structures according to their impact on the earth's gravitational field.

gravity drainage drive: A term given to the extraction of hydrocarbons with the aid of reservoir dip factors, and the resulting force of gravity on fluids within that reservoir.

grease: A semi-solid emulsion of base oil and soap, commonly used as a lubricant.

Greek Fire: An ancient Roman naval terror weapon widely believed to be made from naturally occurring crude oil surface seeps. Greek Fire has been called the ancient precursor to modern napalm.

green coke: An umbrella term for raw petroleum coke, created at a refinery's coker unit, that needs to be calcined to remove excess water and light hydrocarbon contents.

gross domestic product: The sum total of all goods and services produced within a specified timeframe. It is commonly measured per country, but can be measured at the state or regional level.

gross pay: The average vertical height across the entire reservoir, regardless of where a wellbore is drilled.

guide shoe: A downhole device attached to the bottom of a casing string. It is typically shaped with rounded edges and made of a durable material to protect the casing string from subsurface rocks as it is lowered into position for cementing.

H

H_2S splitting: *see Claus Process*

heavy distillate: The largest, most complex hydrocarbon molecules with the highest boiling points. Products in this fraction include residual fuel oil, bitumen and base oils.

heavy fuel oil: *see heavy gas oil*

heavy gas oil: Any liquid, heavy distillates suitable for blending into residual fuel oil.

heavy naphtha: A sub-classification for naphtha-range hydrocarbon molecules with boiling range between 80-90°C and 150-200°C. Heavy naphtha is catalytically reformed into a high-octane gasoline blendstock called reformate.

hoghead: *see mud engineer*

home heating oil (HHO): A middle distillate commonly used as a fuel source for space heaters and cooking equipment. Similar to diesel fuel in hydrocarbon composition, HHO is also referred to as No. 2 Fuel Oil.

hook: Component of a drilling rig's traveling block with which the kelly or other rotation system actually connects to the derrick.

horsehead: Steel component at one end of a pumpjack that ensures straight vertical movement of its downhole lifting equipment.

hot mix asphalt: A mixture of bitumen with aggregates requiring high temperatures, around 250-325°F, to be properly workable for road construction.

Hubbert's Curve Theory: An economic theory that predicts oil production - whether at a field, national, or global level - will follow a roughly bell-shaped curve. Commonly known as Peak Oil Theory or simply Peak Oil.

hydraulic fracturing: A type of well stimulation/completion method in which fluids and proppants are injected into a reservoir under high pressure to increase reservoir porosity and permeability.

hydraulic lift pump: An artificial lift system that uses water or other fluids to power a downhole pump.

hydrocarbon: An organic molecule consisting of both hydrogen and carbon. Commonly used in the industry and throughout this book as a synonym for "oil and natural gas," collectively.

hydrocarbon window: A range of pressure and temperature conditions required for geological forces to create oil and/or natural gas out of decomposed organic matter.

hydrocracking: A type of cracking process that uses pressurized hydrogen gas and high temperatures to break large hydrocarbons into smaller ones.

hydrodesulfurization: An additional treatment refining process in which free hydrogen is bonded to sulfur or other contaminants present within a stream of middle or heavy distillates.

hydroelectric power: Utility-scale electricity generation from the flow of water over a distance.

hydrogen fuel cell vehicles (FCVs): A vehicle powered by a fuel cell, which is device that converts chemical energy into mechanical energy by transferring ions through a membrane in a manner somewhat similar to a battery.

hydrometer: A graduated, glass tool used to measure the density of a liquid substance, including crude oil.

hydrophobic: The tendency of certain molecules to repel water. This physical property allows certain crude oil refined products, including several derived from base oils, to make effective cosmetic moisturizers.

hydroprocessing: *see hydrotreating*

hydrotreating: An additional treatment refining processes in which free hydrogen is bonded to sulfur or other contaminants present within a stream of light distillates, dewaxed oil or natural gas.

I

ICE: *see Intercontinental Exchange*

IEA: *see International Energy Agency*

IFO: *see intermediate fuel oils*

in-solution gas drive: A gas cap drive technique in which natural gas dissolved within liquid crude oil facilitates the lifting of hydrocarbons to the surface from the bottom of the wellbore.

inclinometer: A downhole tool used during oil drilling to facilitate a straight hole test.

Incoterms: A series of international shipping standards outlining common trade practices accepted by governments and industry worldwide.

independent oil company: *see exploration and production company*

injector well: A well drilled, or an existing well converted, for the purpose of injecting fluids to increase reservoir pressure.

insecticide: Natural or synthetic chemicals designed to deter or kill insects, particularly from farm systems but also commonly used in homes and businesses.

Intercontinental Exchange (ICE): Publicly traded commodities marketplace, including futures and derivatives. Brent light sweet crude, the European oil benchmark, is traded on the London-based ICE Futures Exchange.

intermediate casing: Special segment of casing lowered and cemented to the sides of the borehole to prevent contamination of fluids between different subsurface formations and to protect the structural integrity of the borehole.

intermediate fuel oils (IFO): A classification system for commonly traded marine residual fuel grades. IFO grades are similar to ISO specifications.

intermediate product: This term is used to refer to outputs from any refinery process that are not yet suitable for sale to consumers. Included are the various fractions that result from the basic separation stage.

internal combustion engine: An engine design in which the working fluid (air) is mixed within a combustion chamber, commonly a cylinder, located inside the engine itself.

International Energy Agency (IEA): An international entity under the umbrella of the Organization for Economic Cooperation and Development (OECD), founded in response to the 1973 Arab Oil Embargo. In effect, it is the oil-consuming nations' collective response to OPEC.

International Standards Organization (ISO): A conglomeration of many national standard setting agencies. The ISO develops and distributes technical standards for many industries, including oil and gas.

ISO: *see International Standards Organization*

isomerization: An enhanced refining process in which butane's atoms are rearranged from a straight-chained (normal butane) to a branch-chained (isobutane) structure, typically with the aid of a catalyst.

isooctane (C_8H_{18}): IUPAC Name: 2,2,4 - Trimethylpentane, this organic compound is notable as being the 100 standard during an octane rating test.

J

jet engine: A type of internal combustion engine in which air first enters through the front of the plane, is compressed by a series of turbines, mixed with vaporized fuel, ignited and expelled out the rear-pointing nozzle, creating thrust.

jet fuel: A refined crude oil product suitable for combustion within a jet engine, consisting of either a wide or a narrow range of hydrocarbon molecules, depending on how it is manufactured and its intended uses.

K

K Factor: An arbitrary classification system used in a crude oil assay to describe a tested sample as being primarily paraffinic or aromatic. Also called the UOP characterization factor.

kelly: Component of a drilling rig's rotation system, the kelly is a hexagonal, hollow metal pipe that connects the swivel to the drill string through the rotary table.

kerogen: The organic source of crude oil formed from decaying plant and animal life that has combined and compressed over a long period of time. Kerogen can be directly extracted in what is called shale oil.

kerosene: A middle distillate created from crude oil refining used as a space heating and cooking fuel, or as a light source. Kerosene-range hydrocarbons are more commonly used in jet fuel.

kinematic viscosity: A derived viscosity measurement based on the fluid's density, commonly measured in strokes and expressed at a specified reference temperature.

knocking: An automotive malfunction characterized by fuel igniting at the incorrect time, or piston position, within a cylinder. Also called detonation, this can result in major engine damage. A fuel's resistance to this phenomenon is quantified by its octane rating.

L

laterals: A type of well stimulation/completion technique in which a flexible hose or pipe is lowered down the borehole from which a stream of fluid is ejected under high pressure. If aimed properly, the fluid should open new channels in the rock, allowing hydrocarbons to reach the wellbore.

lift (of hydrocarbons): The vertical movement of hydrocarbons to the surface from the wellbore. Reservoir pressure must be significantly higher than surface pressure for this to occur naturally, or else some type of pumping equipment will be needed at the surface of the well.

light distillate: The range of hydrocarbon molecules - and resulting consumer products - with the lowest density and boiling points that are able to condense within an atmospheric distillation unit. This generally includes natural gas liquids, naphtha, and gasoline blendstocks.

light fuel oil: A refined crude oil product classified as a middle distillate and generally used as a space heating and cooking fuel. Commonly referred to as home heating oil.

light gas oil: *see light fuel oil*

light naphtha: A sub-classification for naphtha-range molecules with a boiling point between 15-30°C and 80-90°C generally used as a petrochemical feedstock, or as a low-octane gasoline blendstock.

line drive water flooding: *see dip water flooding*

liquid petroleum gas (LPG): A crude oil and natural gas refined product commonly used as cooking or transportation fuel containing various forms of propane, butane, and trace amounts of ethane and propylene.

LPG: *see liquid petroleum gas*

lubricant: A substance used to reduce friction between solid moving parts, such as grease and motor oil. Many consumer lubricants are made from crude oil refining, particularly base oils.

M

magnetic survey: A hydrocarbon exploration technique that measures differences in magnetic fields caused by the minerals and substances in the earth's crust, including within subsurface rock formations.

Majors: The largest privately owned oil companies in the world, mostly formed through mergers and acquisitions between the so-called Seven Sisters from the industry's oil years. At present the Majors are: BP Plc, Chevron, ExxonMobil, Royal Dutch Shell and Total SA.

making pipe: Nickname used by oil drilling crews for the process of connecting a new segment of pipe to the drill string.

marine diesel fuel: A type of bunker fuel commonly utilized by low speed auxiliary engines on ships for tight navigation through bays, channels and ports.

marine terminal: Strategically located shipping ports acting as major oil trade hubs. It is also common for large quantities of crude oil to be stored at these locations.

maximum reservoir contact well: A type of advanced completion technique in which multiple directional holes are drilled in connection with a single vertical wellbore.

mega-well: A single oil well whose production is so large that it alters the fundamentals of the regional or global market.

membrane filtration: The use of synthetic membranes to sort a mixture of gases according to their molecular size.

mercaptan: *see thiol*

methane (CH_4) (C1): The simplest hydrocarbon and principal component of natural gas, methane is used as a fuel source for electricity generation and, increasingly, transportation.

methane hydrate: A crystal-like substance consisting of methane enclosed within, but not chemically bonded to, frozen water.

methanol (CH_3OH): The simplest type of alcohol, methanol is actually a toxic liquid used primarily as a petrochemical feedstock and to denature fuel ethanol.

methyl tertiary-butyl ether ($C_5H_{12}O$) (MTBE): An organic compound formerly used as a gasoline oxygenate in the United States. This role has largely been replaced by fuel ethanol.

microcrystalline wax: A semi solid crude oil refining product used in the manufacturing of cosmetics having a higher isoparaffin content than paraffin wax.

middle distillate: A range of hydrocarbon molecules - and resulting consumer products - with approximate boiling points between 175-350°C. Included in this category are kerosene, jet fuel, diesel fuel and home heating oil, to name a few.

midstream sector: Refers to the transportation and, by some industry experts, storage of oil and natural gas.

mineral lease: A legal document which conveys the right to extract minerals, including oil and natural gas, from underneath the subject property.

mineral oil: A crude oil refined product high in paraffin content used in a wide variety of consumer products created from dewaxed base oils. This substance is also referred to as white oil and paraffin oil.

molecule: Atoms connected by chemical bonds.

Motor Octane Number (MON): A method of calculating gasoline's octane rating under laboratory conditions by comparing its knock properties to isooctane and n-heptane. MON differs from research octane number to the extent that it uses more vigorous testing conditions.

monkeyboards: Industry jargon for a special platform attached to the top of a drilling derrick used to safely store stands of drill pipe during the process of tripping pipe.

monomer: Unsaturated olefins produced from steam cracking natural gas liquids such as ethane an propane. An input in the manufacturing of plastics.

MON: *see Motor Octane Number*

motor oil: A finished product of crude oil refining used to lubricate automobile engines. Consumer motor oil formulations are typically graded by their viscosity performance at various temperatures.

mousehole: An auxiliary hole through the rig floor and drilled a few feet into the ground in which the next pipe segment is temporarily stored until it is added to the drill string.

MTBE: *see methyl tertiary-butyl ether*

mud: *see drilling fluid*

mud engineer: Specialist in the drilling process tasked with controlling the consistency and composition of drilling fluid as the bit drives deeper into the borehole. Nicknamed 'hoghead' by the drilling crew.

mud pump: The workhouse of the drilling rig's fluid circulation system. The mud pump is typically powered by a diesel engine.

mudlogger: Worker at the drilling rig site who analyzes rock cuttings and drilling fluid returned to the surface for information about the rock structures encountered by the drill bit.

multiple zone production: A well completion technique in which at least two subsurface formations are tapped for hydrocarbons with a single vertical wellbore.

N

n-heptane (C_7H_{16}): Paraffinic hydrocarbon notable for being arbitrarily used as the 0-point in laboratory octane rating testing.

naphtha: A mixture of liquid hydrocarbons, typically with between 6 and 12 carbon atoms each and a boiling range of 30-200°C. Primarily used as a gasoline blendstock and for the manufacturing of petrochemicals and solvents.

naphtha reforming: *see reforming*

narrow cut jet fuel: A crude oil refined product primarily consisting of kerosene-range molecules. This is the most common type of jet fuel and includes Jet A grade for civilian use.

nationalization: The process of seizing private assets into government ownership, either with or without compensation to the former owners.

natural bitumen: *see oil sands*

natural gas: A gaseous mixture of hydrocarbons, including methane, ethane, propane, butane and, in its naturally occurring form beneath the surface, other trace contaminants, including carbon dioxide, sulfur, water, nitrogen and others. Confusingly, this term also refers to the refined consumer product, which consists of almost entirely methane, delivered by pipeline to residents and businesses as a fuel for space heaters, stoves, water heaters etc.

natural gas drive: *see gas cap drive*

natural gas liquids (NGLs): That portion of natural gas which exists as a liquid at surface temperature and pressure, or is otherwise considered by the industry as "easily condensable." Included in this category are ethane, propane, butane, pentane and larger molecules.

natural gas processing plant: Equipment and infrastructure dedicated to the purification of natural gas. This includes sulfur, water and contaminant removal, as well as the isolation of the various hydrocarbon molecules that make up raw natural gas.

natural gas vehicle (NGV): Vehicles powered by compressed natural gas. In the U.S., most NGVs are bus and government fleets. Not to be confused with vehicles that run on LPG (propane).

natural lift: *see flush production*

Nelson Complexity index: An arbitrary method of analyzing a refinery's complexity, or the degree to which it can process heavy (and/or sour) crude blends.

netback pricing: A crude oil pricing mechanism which takes into account the producer's location by, for example, factoring shipping costs into the wellhead price.

neutralization number: *see total acid number*

New York Mercantile Exchange (NYMEX): Publicly traded commodities marketplace, headquartered in New York, in which crude oil, natural gas, various refined products and other commodities are traded.

NGLs: *see natural gas liquid*

NGV: *see natural gas vehicle*

NIMBY: "Not in My Back Yard." Famous example: many constituencies throughout the country are in favor of the government building more prisons, as long as they are located far away from their homes and schools and businesses.

non-associated gas: Natural gas produced without crude oil by, for example, a pure natural gas well.

nonattainment area: Regions of the U.S. designated by the EPA as having poor air quality according to the standards of the Clean Air Act. Such regions have differing fuel standards for automobiles, including the need for reformulated gasoline.

nuclear power: Utility-scale electricity generation from a controlled, sustained chemical reaction called fission.

NYMEX: *see New York Mercantile Exchange*

O

Occupational Safety and Health Administration (OSHA): Federal government agency under the U.S. Department of Labor tasked with assuring safe working conditions. Included activities under their jurisdiction are oil and gas drilling and crude oil refining.

octane rating: Derived from a variety of laboratory techniques, octane rating essentially quantifies a tested gasoline's resistance to knocking, an important property of the fuel.

off-spec crude: Crude oil that is unmarketable to any nearby refineries. This can be due to the oil having too much sulfur, acid or water content, or too high a density. Transportation costs from the wellhead to the refinery are also a factor.

offgas: Natural gas molecules released from a fluid catalytic cracker's re-distillation chamber.

offshore drilling: The process of making a borehole through the bottom of the ocean, generally accomplished with a drilling rig on some type of ship.

oil: A complex mixture of liquid hydrocarbons generally thought to be created from compressed organic matter over a long period of time. Conventionally extracted from the pores of subsurface rock formations.

oil embargo: Similar to the concept of a boycott. This can refer either to exporters refusing to sell (e.g. the 1973 Arab Oil Embargo) or buyers refusing to purchase oil from one or more countries (e.g. the 2011 U.S. and EU trade embargo on Iran).

oil field: This term is commonly used to refer to the surface area above a single subsurface oil-bearing reservoir.

oil formation: A subsurface rock formation that contains liquid hydrocarbons. It must also be of sufficient porosity and permeability for hydrocarbon extraction to be possible.

oil sands (natural bitumen): Higher-density hydrocarbons combined with loose sand. Before it can be processed by conventional refineries, sand and oil must be separated and, in certain cases, the oil must be upgraded to be suitable for transport through pipelines.

oil shale: Kerogen-rich source rock of a shale-like structure. Can be strip-mined in a manner similar to coal, and the resulting shale oil can be refined into products like gasoline and diesel fuel. Often mistakenly confused with the concept of "tight oil."

oil well: A vertical or directional hole in the ground drilled specifically to extract liquid crude oil.

oil-wet pore: When liquid crude oil is essentially "between" its solid rock reservoir and any water that may be present within the pore, that pore is said to be oil-wet.

olefin: The traditional oil and gas term for a what is properly called an alkene, which is an unsaturated branch-chained hydrocarbon. Olefins are created artificially through a variety of refining processes, particularly steam cracking.

online: A well is said to be "brought online" when it starts producing, either for the first time or after being shut in for any length of time.

OOIP: *see original oil in place*

OPEC: *see Organization of Petroleum Exporting Countries*

open-hole completion: A type of completion in which no production casing is set through the oil-bearing formation's rock layer. This allows fluids to directly enter the borehole.

operating agreement: An agreement between the owner(s) of a mineral lease and the business entity tasked with the actual drilling, completion and, hopefully, production phases of one or more wells.

operator: The entity whom, by agreement, has decision-making authority and responsibility for the various day-to-day tasks related to bringing a new well online and, if applicable, managing the production life cycle of one or more wells in a field.

Organization of Economic Cooperation and Development (OECD): An economic bloc of nations ultimately arisen out of the Marshall Plan. Now consisting of 34 member countries, the OECD advocates free trade, global growth and democracy, among other topics.

Organization of Petroleum Exporting Countries (OPEC): An economic bloc of oil-exporting nations created in 1960 in an attempt to coordinate its members' oil producing activities, including exports. Using production quotas, OPEC influences the global oil market by restricting or increasing supply.

original oil in place (OOIP): An oil reserves term that refers to all liquid hydrocarbons in place within a reservoir, with no distinction of whether they are commercially viable or technically feasible to extract.

OSHA: *see Occupational Safety and Health Administration*

oxygenate: Gasoline additives which increase the oxygen content of the fuel, providing a more thorough, clean, environmentally friendly burn. In the U.S., this is commonly achieved with the use of ethanol fuel.

P

P&A: *see plug and abandon*

packer: A downhole device which is lowered into the borehole and then expands, essentially acting as a plug. This device is an essential requisite for certain completion tasks, such as perforation and drill stem testing.

packing: The use of special materials to increase the surface contact between various hydrocarbon molecules within a vacuum distillation chamber instead of the horizontal trays used in an atmospheric distillation unit.

PADD: *see Petroleum Administration for Defense Districts*

PAOs: *see polyalphaolefins*

paraffin: The traditional oil and gas term for what is properly called an alkane, which is a straight-chain, saturated hydrocarbon commonly found in liquid crude oil.

paraffin oil: *see mineral oil*

paraffin wax: A highly viscous, semi-solid crude oil refined product consisting of straight-chained alkanes. Paraffin wax is used to make candles and a variety of "food grade" coatings and other products.

payzone: An oil-bearing rock formation having physical properties, including porosity and permeability, suitable for commercially viable extraction.

peaker plant: Supplemental electricity generated during peak demand, such as during hot summer days. Certain energy sources are more suitable for this purpose, e.g. natural gas, while others are less, e.g. nuclear and hydroelectric power.

penetration rating: A method of classifying the hardness of solid or semi-solid petroleum products like bitumen and wax.

pentanes plus: An umbrella term for all hydrocarbons within natural gas larger than (and not including) butane.

percussion coring: *see sidewall core sampling*

perforation gun: A device lowered into the borehole by wireline from which several small explosive charges are shot through the production casing and into the rock, allowing reservoir fluids to enter the wellbore.

perforation: A type of well completion method in which small explosive charges are fired through the production casing and cement into the target formation.

permeability: The measurable tendency of fluids to flow through a network of connected pores or void spaces within a solid rock structure.

petrochemical: Any of a wide variety of substances created from crude oil refining commonly used as a feedstock in the creation of numerous finished products, including plastics, solvents, fertilizers, pesticides, and many others.

petrodiesel: *see diesel fuel*

petrol: *see gasoline*

petrolatum: *see petroleum jelly*

petroleum: *see oil*

Petroleum Administration for Defense Districts (PADD): A bureaucratic organization of the 50 states into five geographic regions for the purposes of aggregating and presenting data about the imports, exports, production, refining, distribution and consumption of crude oil, natural gas, and many of their various products by the Energy Information Administration (EIA)

petroleum coke: A refined crude oil product high in carbon content used as an industrial fuel in a manner similar to coal.

petroleum gas: *see natural gas*

petroleum jelly: A refined crude oil product derived from the heavy end of the barrel. As with some other base oils, it is a semi-solid at standard temperature and pressure. Famously traded as Vaseline.

petroleum-based products: A catch-all term for any of a wide variety of consumer, commercial and industrial products derived from oil, as well as those made from natural gas.

pick-up ticket: A receipt for the sale of oil by a single well or field to a refinery or distribution company.

pin: The "male" threaded end of a segment of drill pipe's tool joint.

pipeline: A conduit through which crude oil, its finished products or, under some compression, natural gas is transferred by and between various market participants.

pipeline inspection gauge ("pig"): A device that travels through an operating pipeline, either for the purpose of cleaning its inner diameter or to isolate batches of product, such as in between a flow of gasoline and diesel fuel which are obviously not meant to be mixed.

pitch: A highly viscous semi-solid produced from the dry distillation of wood. Often confused with natural bitumen, a high-density hydrocarbon substance found in oil sands.

plastic: Refers to any of variety of organic polymers easily capable of being molded or cast into custom shapes. Many plastics are ultimately derived from hydrocarbons like ethane and propane.

plug and abandon (P&A): A process for sealing off a drilled well by injecting cement and mud to prevent communication of fluids between its subsurface formations. This is performed on both dry holes and after a well's reserves are depleted.

plunger lift pump: A relatively simple natural lift system that drops a plunger to the bottom of the borehole, closing a downhole valve. After some time, enough pressure builds up at the bottom of the borehole, at which time the plunger rises to the surface, pushing up a quantity of liquid hydrocarbons.

polyalphaolefins (PAOs): A type of polymerized alkene classified as a Group IV synthetic base oil according to the API's classification of motor vehicle oils.

polymer: A single molecule consisting of a large monomer repeated many times. The ability to create such a structure is the foundation of the plastics industry, although natural polymers also exist.

polymerization: An enhancement refining technique that combines smaller hydrocarbon molecules (natural gas) into larger, more complex ones that can be more readily used as transportation fuel.

pores (of a rock): Open space within a rock matrix in which water, natural gas, crude oil or other compounds might naturally accumulate.

porosity: A ratio of the total volume of a sample of rock's pores to its total volume. This is one of the most important physical properties when evaluating a potential oil formation's commercial viability.

possible plus probable plus proven (P10) (3P): A cumulative estimate of reserves that might be achieved, but only under favorable circumstances, indicating a 10% degree of confidence.

possible reserves (P3): Those reserves which are given a probability estimate of 10-49.9% of existing within a reservoir, region or nation.

posted price: A refinery's published starting point for negotiations between the company and producers for the purchase of crude oil according to its quality.

pour point: The temperature below which a liquid will not pour, acting more like a semi-solid instead.

prechamber: An indirect injection method whereby combustion occurs in a different component of the cylinder. This type of system is better suited to the usage of straight vegetable oil as fuel, and for improved cold engine starting.

primary recovery: For the purposes of this book, the term refers to the stage of a well's life cycle in which reservoir pressure is not sufficient to vertically lift hydrocarbons to the surface, requiring installation of some type of pumping equipment.

probable reserves (P2): Those reserves which are given a probability estimate of 50-89.9% of existing within a reservoir, region or nation

processing gain: *see refinery gain*

produced water: The water that is produced along with oil or natural gas at the wellhead. A well can produce many times more water than oil, and this water is typically separated at the field where it is produced.

production capacity: The maximum quantity of output that a well, field, region or nation is capable of producing under ideal conditions over some unit of time, usually expressed as barrels per day (bpd).

production casing: A segment of case that typically runs from the surface to the bottom of the well which will be tapped for hydrocarbons.

production phase: That stage of a well's life cycle during which hydrocarbons are extracted from a reservoir over a long period of time, often for many years. It includes the stages of flush production, primary recovery and secondary recovery, and ends when the well is depleted.

production quota: Formerly utilized by the Texas Railroad Commission and now used by OPEC, a production quota is a maximum allowable quantity of crude oil that a participant can bring to market. Such quotas are ultimately used to control prices by restricting, or increasing, supply as needed.

production tubing: An additional segment of hollow, metal pipe lowered through the borehole to the target zone, protecting the actual casing from corrosion that might be caused by the various contaminants present within crude oil.

productivity index: A reservoir analysis measurement calculated by dividing a well or field's daily production rate by its reservoir drawdown pressure.

progressive cavity pump: An artificial lift system in which fluids are brought to the surface by means of a rotating helical space.

propane (C_3H_8) (C3): A gaseous paraffinic hydrocarbon and component of natural gas. Propane is used as a petrochemical feedstock, a transportation fuel and, as it most commonly known by the public, cooking fuel.

proppant: A solid material used in the hydraulic fracturing process to hold open newly created void spaces and channels, facilitating the flow of hydrocarbons through the reservoir to the wellbore.

proven plus probable (P50) (2P): A cumulative estimate of reserves that is *as likely as not* to be achieved, indicating a 50% degree of confidence.

proven reserves (P1) (P90) (1P): Those reserves which are given a probability estimate of at least 90% of existing within a reservoir, region or nation

pulley: A mechanical device that supports the lifting of weight. Adding more wheels to the system distributes the work in such a way that less force is needed to lift the weight off the ground. This is the fundamental component of a drilling rig.

pump octane number (PON): *see anti-knock index*

pumparound: Component of an atmospheric distillation unit that returns the reflux, or higher density hydrocarbons not drawn off as pure distillate, back into the distillation chamber.

pumpjack: *see sucker rod lift pump*

pyrolysis: A type of combustion in which heat is applied to a substance in the absence of oxygen.

R

rack: A framework of metal grooves on the rig floor which hold the bottoms of drilling pipe stands during the tripping pipe process along with the fingers of the monkeyboard at the top.

radiometric survey: A hydrocarbon exploration technique that measures an area's subsurface natural radioactivity. Hydrocarbons are known to disrupt this pattern, if present.

raffinate: An intermediate crude oil product used as a feedstock in base oil, grease and wax manufacturing, and as a low octane, low RVP gasoline blendstock..

rathole: An auxiliary hole through the rig floor and drilled a few feet into the ground in which the kelly unit is stored while the crew is making or tripping pipe.

RBOB: *see reformulated blendstock for oxygenate blending*

redistillation chamber: An enhanced refining device which sorts and separates hydrocarbons, according to density, that have just gone through a fluid catalytic cracking (FCC) unit.

recompletion: Any of a number of well stimulation/completion techniques applied to a well that has already been producing for some time. Also sometimes called reworks, these generally involve attempts to increase the lateral movement of hydrocarbons across a reservoir.

recovery efficiency: The percentage of a reservoir's Original Oil in Place which are considered technically recoverable under current circumstances.

refinery: A facility which carries out the processes and procedures of converting raw crude oil and natural gas into usable consumer-ready substances, including a variety of fuels, lubricants and petrochemicals collectively referred to as petroleum-based products.

refinery complexity: This downstream market analysis metric is essentially an attempt to quantify, with simple numbers, the ability of a refinery, company or entire nation to profitably process high density oil inputs.

refinery gain: The difference in volume between what a refinery gets out of a crude barrel and the 42 gallons that make up that barrel.

reflux: Once captured from the atmospheric distillation unit's stripper tanks, higher density hydrocarbons belonging to a lower tray level are separated from a purified distillate product and reinserted into the chamber via the pumparound.

reforming: An enhanced refining process in which naphtha-range molecules are chemically rearranged into those with notably higher octane rating, suitable for blending into gasoline.

reformulated blendstock for oxygenate blending (RBOB): A publicly traded commodity, RBOB is simply RFG-grade gasoline that hasn't yet had its oxygenate content added.

reformulated gasoline (RFG): An alternate grade of federally approved finished motor gasoline, RFG is required by the EPA's Clean Air Act to meet tougher pollution control standards than conventional gasoline. These standards are then generally met with higher oxygenates content.

regenerator: Component of a fluid catalytic cracking unit in which accumulated carbon deposits are burned off the zeolite catalyst.

reheater: Also called a reboiler, this external device draws off liquid oil from the bottom of the atmospheric distillation unit and reapplies heat. Any vapors released from this device are then reinserted into the chamber to be separated according to density.

Reid Vapor Pressure: A measure of the volatility of crude oil and its associated finished products. A more volatile product will more quickly evaporate into the atmosphere.

renewable energy: Renewable energy refers to those sources of alternative energy which also come from renewable sources, including wind, solar, geothermal and biomass. Nuclear power is alternative, but not renewable.

Research Octane Number (RON): A method of calculating gasoline's octane rating under laboratory conditions by comparing its knock properties to those of isooctane and n-heptane. RON results differ from Motor Octane Number because it uses less vigorous testing conditions.

reserves: An estimate of a quantity of hydrocarbons which are both technically feasible and economically viable to extract.

reservoir: That portion of a subsurface, hydrocarbon-bearing rock formation which contains oil or natural gas. Can be used interchangeably with the term oil formation.

reservoir analysis phase: Any of a number of analysis and measurement techniques performed during or after the actual drilling of the well to determine whether the new well will be commercially viable enough to warrant the completion phase.

reservoir areal extent: The surface area above an oil reservoir. Similar to a concept of an oil field, this number is used in the volumetric method of estimated original oil in place (OOIP).

reservoir pressure: The average pressure within the reservoir. If substantially higher than standard surface pressure, hydrocarbons may be pushed toward the surface.

reservoir vertical extent: Also called 'net pay,' this is the vertical height of the target formation at the wellbore.

residual fuel oil: A refined crude oil product classified as a heavy distillate generally used as an industrial, commercial and international shipping fuel.

resources: Quantity of estimated oil or natural gas that is technically recoverable. Unlike reserves, economic viability is not a factor in this classification.

return on investment (ROI): One of the most widely used investment analysis methods, particularly useful at comparing different opportunities to each other. It is calculated by dividing the net profit of an investment by its cost.

rework: The process of repairing existing completion equipment on an online or shut in well or field, such as the replacement of production tubing or repair of an artificial lift system.

RFG: *see reformulated gasoline*

riding valve: *see ball and seat valve*

rig floor: A manmade platform located a few feet above the intended borehole on which the drilling crew performs many of its tasks, including making and tripping pipe.

rig superintendent: The highest ranking individual on the drilling rig crew. Nicknamed toolpusher, this individual ensures crew safety and works closely with the operator throughout the drilling process.

riser: Component of a fluid catalytic cracking unit in which vaporized hydrocarbons interact with the zeolite catalyst and cracking occurs.

risk premium: That portion of the open-market oil price directly attributed to global security concerns, such as supply disruptions, terrorist attacks, even natural disasters.

roller cone drill bit: A type of rotary drill bit with independent steel cones connected via a bearing system. The cones, in turn, each have teeth or inserts for the cutting and grinding of subsurface rock.

RON: *see Research Octane Number*

rotary drilling: The conventional method of drilling a borehole into the ground by which a bit and a string of pipe segments are rotated by a variety of possible methods. The rotating bit then cuts through the rock, with downward force applied by the pipe above it.

rotary hose: A length of heavy-duty, flexible tubing which connects the mud pump to the top of the drill string.

rotor: The part of a generator that rotates within the non-moving stator. Also typically has alternating magnetic fields attached.

roughnecks: Industry term for the crew which operates the rig throughout the drilling process.

roustabouts: Newer members of a drilling crew generally responsible for maintaining the rig and its equipment, and other entry-level tasks.

S

SAE International: U.S.-based trade organization that develops technical standards for automobiles and automotive products, such as greases and motor oils.

samples: *see cuttings*

saponification: A special type of chemical reaction used, among other things, to make soap.

saturation water percentage (SW): The ratio of a reservoir's water content to its total fluid volume.

SCO: *see synthetic crude oil*

scratchers: Mechanical devices wrapped around casing as it is lowered into the borehole that scrape accumulated mud deposits, allowing the cement to bond casing directly to borehole.

SEC: *see Securities and Exchange Commission*

secondary recovery: For the purposes of this book, the term refers to the stage of a well's life cycle in which reservoir pressure is not sufficient to enough to readily draw hydrocarbons laterally across the reservoir to the wellbore, instead requiring enhanced techniques such as recompletions and water flooding.

Securities and Exchange Commission (SEC): A U.S. federal agency tasked with regulating various markets, including stocks, bonds, and many other investment vehicles.

seep: A natural occurrence of hydrocarbons reaching the surface due to the lack of any impermeable cap rock layer above a reservoir. Reservoir pressure and buoyancy combine to propel oil and gas to the surface. This phenomenon has been known about and exploited for centuries.

seismic survey: A hydrocarbon exploration technique that measures the way artificially created sound waves interact with various layers of subsurface rock and sediment around a potential oil field.

shale gas: Natural gas extracted from a shale-like rock structure. Due to the properties of shale formations, this scenario was not considered viable prior to the advent and perfection of modern-day hydraulic fracturing and horizontal drilling techniques.

shale oil: Extremely high density hydrocarbon mixture called kerogen, a geologic precursor to crude oil, produced from a special type of sedimentary rock matrix called shale. Shale oil is strip mined and is therefore considered an unconventional petroleum resource.

shear: In mechanical engineering, a shear force is one that attempts to cause the internal structure of a material to deform.

sheaves: *see pulley*

shut in: Refers to an oil or gas well being offline, or not producing. This could be temporarily, such as for emergency repairs or scheduled maintenance, or for an extended period of time, usually due to market factors.

sidewall core sampling: A reservoir analysis technique whereby small pieces of reservoir rock are extracted by means of a wireline tool lowered into the borehole. Small, hollow cylinders are detonated from this tool while connected to a wire, which are later pulled back to the surface.

sinker bar: A heavy weight used by a cable-tool drilling rig to weigh down the chisel bit. In modern drilling, sinker bars are still used to weigh down various wireline tools.

slips: A tool used by a drilling rig crew to suspend the weight of the drill string while the kelly and derrick are disconnected when making or tripping pipe.

slip puller: Member of the rig crew tasked with operating the slips when making or breaking pipe.

smoke point: The operating temperature of an engine, usually a jet engine, at which fuel starts to burn fuel inefficiently.

Society of Automotive Engineers: *see SAE International*

solar panel: A device which converts photons emitted by the sun into electric current through what is called the photovoltaic effect.

solar power: Utility-scale electricity generation derived from the sun's light, heat or both. Individual households or buildings with rooftop solar panels are usually classified under the broader term "solar energy."

solid-desiccant dehydration: A method of removing water from a natural gas stream, typically within a pipeline, by means of adsorption facilitated by a granular metallic compound.

solution gas-oil ratio: A reservoir analysis metric which estimates the amount of natural gas that will leave the liquid-gas solution per each barrel of liquid oil.

solvent: A substance into which other substances can be dissolved and, usually, later extracted.

sour crude: Refers to an oil blend which contains greater than or equal to 1.5% sulfur content (15,000ppm) by weight.

source rock: A subsurface layer of compressed organic matter that, upon further exposure to temperatures and pressure over time, may generate oil and natural gas.

space heating: The use of fuel or electricity to heat an enclosed space, such as a home or some other building. During the winter months in the U.S., space heating demand causes increased demand for many crude oil products.

specific gravity: A measure of relative density, specific gravity is the ratio of a sample's density to that of water.

speculation: The practice of purchasing futures contracts for oil, consumer gasoline or other petroleum products solely on the expectation of higher nominal prices in the future, whether due to supply issues, geopolitical events or similar reasons.

Spindletop: One of the world's first "gushers," this mega-well was drilled in 1901 and singlehandedly "put America on the map" as a global oil power.

spot pricing: An exchange-traded or over-the-counter contract for the immediate purchase of a commodity at a specific location. This pricing mechanism is more volatile than futures pricing.

SPR: *see Strategic Petroleum Reserve*

spudding: The process of breaking ground on a new borehole or the drilling shallow auxiliary holes, such as the rathole or mousehole.

stability: The measurement of the degree to which a volatile substance, such as fuels, can be stored safely and without losing their combustible properties.

stack: In the energy industry, the word "stack" refers to the variety of energy sources a utility company has available at any given time, and are generally said to be sorted by their desirability as a fuel.

stamping: *see whole core sampling*

stand: A collection of drill pipe segments, usually between two and four, which remain connected during the process of tripping price, stored between the rack and the monkeyboards.

Standard Oil Company: Founded in 1870 by John D. Rockefeller, Standard Oil dominated the U.S. and global industry for decades until it was broken up by anti-trust regulators.

standing valve: *see ball and seat valve*

stator: The non-moving part of a generator that houses the spinning rotor inside. The stator is wrapped in metal coils, usually made of copper.

steam cracking: The use of extremely high temperature steam to crack hydrocarbons, primarily used to create olefins as petrochemical feedstocks.

step-out: The practice of strategically placing additional production wells directly adjacent to a previously successful one.

straight hole test: An importance diagnostic performed during drilling to ensure that the borehole is drilled with as little deviation as possible.

straight run bitumen: *see straight run residue*

straight run distillate: Hydrocarbons in the middle distillate-range coming out of the basic separation stage of an oil refinery. SR distillates are generally used as inputs into the various enhanced refining processes, but are sometimes directly blended into finished products like kerosene, jet fuel, and diesel fuel.

straight run gasoline: *see heavy naphtha*

straight run residue: Liquid hydrocarbons which are generally not distilled during the basic separation stage of oil refining. Used as a feedstock for a variety of enhanced refining processes. Sometimes referred to as vacuum gasoil when bought and sold between refineries in lieu of further processing.

straight vegetable oil (SVO): A fuel source derived from used cooking oil, such as the oil used by a restaurant's deep fryers. Unlike biodiesel, which is processed somewhat, SVO fuel must be preheated prior to combustion in a diesel engine.

Strategic Petroleum Reserve (SPR): A collection of facilities maintained by the U.S. government to stockpile quantities of crude oil in the event of a major supply disruption, such as war or natural disasters. The SPR was established in 1975 in response to the 1973 Arab Oil Embargo.

stratigraphic traps: Any of a variety of hydrocarbon reservoirs trapped by an interrupting, impermeable rock layer, or that occur when the reservoir itself experiences changing physical properties across its rock bed.

stripper tank: Component of an atmospheric distillation unit within which reflux and pure distillate products are separated and sorted.

stroke length: The absolute distance between a pumpjack's location at bottom dead center and top dead center.

structural traps: Any of a variety of hydrocarbon reservoirs trapped due to tectonic events, such as geologic folding or faults.

structure: In the oil drilling industry, the structure provides level ground above the drilling pad. This extra height facilitates the processes of making and tripping pipe.

subsurface safety valve (SSSV): A downhole device connected to the production tubing capable of shutting in the well in the event any surface equipment is destroyed or otherwise fails.

sucker rod lift pump: An artificial lift system in which a plunger is lowered and raised by a metallic rod by means of a reciprocating engine at the surface.

summer blend gasoline: A gasoline variant that contains fewer low density hydrocarbons, such as butane. In the summer, volatile fuels are more likely to evaporate into the air. Evaporated gasoline causes smog.

summer driving season: In the U.S., the months of the year between Memorial Day and Labor Day generally associated with additional vacation travel.

surface casing: The segment of casing, of a length largely determined by local regulations, lowered and cemented in the borehole to protect shallow freshwater resources, and to protect the structural integrity of the borehole.

surfactant: A chemical compound that allows two insoluble liquids to effectively mix by reducing the interfacial tension between them.

swabbing: A completion technique used to remove excess fluids from a wellbore by lowering a wireline tool surrounded by cup-shaped devices. When removed, the cups create a suction force, pulling all fluids up to the surface.

sweet crude: Refers to an oil blend which contains less than 0.5% sulfur content (5000 ppm) by weight.

sweetening: An additional treatment process of crude oil refining in which hazardous sulfur compounds are either removed outright or oxidized into less dangerous forms.

swing producer: An oil producer, whether a country or company, operating at less than full capacity. Idle capacity can be later employed to mitigate rising prices as needed.

swivel: A component of a rotary drilling rig which connects the drill string to the derrick. It allows the drill string to rotate without having to rotate the entire derrick.

synthetic base oil: Highly refined base oils are said to be synthetic due to their higher performance properties than other traditional blends. Nearly all consumer engine oils are synthetic.

synthetic crude oil (SCO): This term has been used to refer to "upgraded" natural bitumen produced from oil sands. Synthetic crude oil is considered a more widely marketable product than natural bitumen as it can be processed by somewhat-lower complexity refineries.

T

tank battery: One or more structures located at an individual well or field that collect and store extracted hydrocarbons, either with or without any produced water.

tar: A byproduct from the pyrolysis of wood, tar was used for waterproofing ships and buildings, and as a medicinal product. The term is often incorrectly used interchangeably with oil sands natural bitumen.

TD: *see total depth*

term contract: A type of futures contract in which a buyer and seller agree to transfer a specified quantity of oil at regular intervals at an agreed upon price.

tetra-ethyl lead ($C_8H_{20}Pb$) (TEL): This compound was used as a very effective anti-knock additive in gasoline until being phased out in the 1970s for a variety of reasons.

Texas Railroad Commission (TRC): State-level government agency tasked with regulating many aspects of Texas' energy industry. Widely considered the precursor to OPEC in its ability to dictate global oil prices by controlling production levels.

thermal cracking: An umbrella term for a variety of enhanced refining processes in which extreme temperature and, in some cases, elevated pressure are used to split large hydrocarbon molecules into smaller ones more suitable for blending into transportation fuel.

thiol: The proper designation for a type of sulfur compound traditionally called mercaptans in the industry. Thiols have strong odor, though not the same as the rotten egg smell of hydrogen sulfide. Trace amounts of thiols are added to many consumer natural gas products as an oderant to aid the detection of gas leaks.

thumper: A vehicle mounted with special seismic equipment used to emit precise, consistent seismic waves into the ground.

tight oil: Refers to relatively low-density crude oil extracted from any shale-like structure. The permeability of shale in general is said to be "tight." This term is very often confused with oil shale, which contains a higher density hydrocarbon substance called kerogen.

tongs: Essentially an oversized wrench, tongs are used to connect or disconnect segments of pipe with the aid of both roughnecks and the rig's catheads.

tool joint: Refers to the threaded male and female ends of a drill pipe segment. To reduce weight relative to drilling collars, tool joints are welded on instead of being directly machined onto the pipe.

toolpusher: *see rig superintendent*

top dead center (TDC): The point in a pumpjack's stroke when the horsehead is the furthest from the ground.

top drive: An alternate rotary drilling system in which a mechanical motor is connected to the traveling block. Replacing the turntable system, the top drive directly rotates the drill kelly and drill string.

total acid number (TAN): A measurement of an oil blend's acidity, TAN is reported as the number of milligrams of potassium hydroxide that would be needed to neutralize a sample of crude oil.

total depth (TD): The intended vertical distance for the well to be drilled. Sometimes called target depth.

trade hub: A location where commodities and other goods are frequently traded. Trade hubs are generally associated with ports, major population centers, or other geographically significant locations.

transesterification: A chemical process utilized in the creation of biodiesel fuel.

transportation fuel: Energy sources used to facilitate transportation (crude oil, natural gas, hydrogen) are commonly distinguished from, for example, electricity-generating resources, such as coal, wind, and solar in the industry.

traveling block: The non-permanent set of sheaves lowered as each new pipe segment is driven down the borehole and raised again when the next segment begins.

tray: A horizontal plate within an atmospheric distillation unit on which condensed hydrocarbons are stored and mixed.

TRC: *see Texas Railroad Commission*

tree: Refers to the actual valves, gauges, and other equipment located at the surface wellhead. Also called the Christmas tree, it controls the flow of fluids from the reservoir to the surface.

tribology: A branch of mechanical and chemical engineering dedicated to the study of friction and lubrication.

tripping pipe: Nickname used by oil drilling crews for the process of quickly removing the drill string, typically to install casing or to lower downhole testing equipment. When removed, two to four segments are typically left connected, called stands.

trunk line: Large diameter pipelines which feed into other pipeline networks or, in the case of crude oil, directly to refining hubs.

turnaround: A period of time during which refineries shut down their processing equipment for maintenance, to replace catalysts, and to retool their operations between so-called distillate mode and gasoline mode.

turntable: A turntable is part of a rotary-type drilling rig. It is connected to the rig floor and mechanically spins the drill string.

U

ULSD: *see ultra low sulfur diesel fuel*

ultimate recoverable resources (URR): *see resources*

ultra low sulfur diesel fuel (ULSD): Refers to any diesel fuel which contains not more than 15 parts per million sulfur. It has been required in the U.S. for on-road vehicles since 2006 and off-road vehicles (including trains and boats) since 2012.

unconventional oil and gas: Any method of extracting hydrocarbons other than by drilling into subsurface formations which have favorable porosity and permeability. Included in this category are those sources which are only economically viable with the use of hydraulic fracturing or horizontal drilling.

undeveloped reserves: A sub-classification of proven, probable, and possible reserves that require significant capital to extract.

United States Geological Survey (USGS): A non-regulatory, scientific federal agency, the USGS is tasked with studying and understanding the country's natural resources, including minerals and water supplies.

upgrading: Any of a variety of processes which convert the high-density semi-solid hydrocarbons extracted from oil sands into medium-density liquids, called synthetic crude oil, more suitable for conventional oil refining.

upstream sector: The processes, facilities, personnel and infrastructure dedicated to the extraction of oil and gas from the ground. Included in this category are the exploration, drilling, completion, and production phases of an oil or gas well.

uranium (U): The heaviest naturally occurring element. Uranium is radioactive, metallic and, when enriched, can be used as a fuel source with the process of nuclear fission.

USGS: *see United States Geological Survey*

utility-scale electricity generation: The term refers to electricity produced for public consumption by utility companies, as opposed to what has been termed distributed generation, which includes rooftop solar panels and household geothermal systems, among others.

V

vacuum distillation: A supplemental basic separation process in which the raw straight run residue from atmospheric distillation is re-distilled under greatly reduced pressure, lowering the boiling point of the feedstock to facilitate its evaporation.

vacuum flashing: *see vacuum distillation*

visbreaking: An enhanced refining process in which heavy distillates are alternatingly exposed to high and low temperatures, cracking the large molecules into smaller ones and reducing the viscosity of the un-crackable heavy molecules that remain.

viscometer: A hollow device through which a liquid is allowed to flow in order to measure its viscosity.

viscosity: A measurable physical property that quantifies a liquid's resistance to flow when interacting with solids or other liquids. The viscosity of crude oil is an indicator of its ability to be refined into lucrative middle and light distillates.

viscosity index: A derived measurement used by the automobile industry to classify lubricants. It better quantifies a substance's performance attributes across a range of operating temperatures.

volatility: A measurable physical property that quantifies a liquid's tendency to vaporize, or become a gas. This measurement is important for many petroleum products because evaporated hydrocarbons are a major contributor to smog.

W

warm mix asphalt: An asphalt product that uses organic waxes or fatty amides to lower the workable temperatures of the substance.

water coning: Occurs when water is pulled to the wellbore more quickly than crude oil, displacing the oil along the permeable channels, resulting in an increase in the output of produced water at the surface.

water encroachment: *see water coning*

water flooding: A secondary recovery technique in which water is strategically injected into a reservoir in order to physically displace any remaining hydrocarbons toward the wellbore for extraction.

water-wet pore: When liquid water is essentially "between" its solid rock reservoir and any liquid crude oil that may be present within the pore, that pore is said to be water-wet.

wax: Any of a variety of crude oil refined products consisting of a solid or semi-solid mixture of hydrocarbons derived from dewaxed raffinate.

wedge: A downhole tool used to facilitate directional drilling. Also called a whipstock, the steel wedge creates a slanted angle for the drill bit to follow. Can also be used to correct undesirable deviation in a straight vertical borehole.

well log: A reservoir analysis technique in which sensors are lowered down the borehole by wireline to study any of a variety of the well's physical properties.

well spacing: The distance between various oil and gas wells. This distance is often regulated by state and federal agencies but is also optimized by production companies according to reservoir characteristics.

well stimulation: *see recompletion*

wellbore: A hole drilled through the ground for the purpose of searching for and, if present, extracting natural resources, including water, oil or natural gas.

wellhead: The surface above an oil or gas well. This includes the tree and any other production equipment that provides pressure control.

West Texas Intermediate (WTI): The American standard benchmark crude oil blend frequently cited by market commentators and analysts. WTI contracts are traded on the New York Mercantile Exchange in New York, and physical delivery occurs in Cushing, Oklahoma.

wet gas: *see condensate*

wettability: As water and oil are insoluble, only one or the other will be in direct contact with the solid rock of a formation. The term wettability refers to whether oil or water is in direct contact with rock within a reservoir.

whale oil: Extracted from whale blubber, whale oil was the primary lighting fuel source for oil lamps throughout the 1800s. Its usage was largely replaced by kerosene in the second half of the 19th century.

white oil: Generic term used to classify light and middle distillates to distinguish them from heavy distillates and raw crude oil itself, especially from an environmental context.

white oil pipeline: *see clean pipeline*

whole core sampling: A reservoir analysis technique whereby a single, large piece of reservoir rock is extracted by means of a special downhole assembly, called a coring barrel, connected to the bottom of the drill string and lowered to the bottom of the borehole.

wide cut jet fuel: A crude oil refined product consisting of both kerosene- and naphtha-range molecules. Due to the increased profitability of allocating naphtha for gasoline, this type of jet fuel is relatively more expensive and less commonly used.

wildcat: The first oil well drilled to search for new hydrocarbon resources in an unproven location. If successful, a wildcat well commonly leads to additional production wells being drilled nearby.

wind power: Utility-scale electricity generation derived from the kinetic energy of wind. This is most often accomplished using wind turbines.

wind turbine: A device which generates wind power by means of large blades that are rotated around an axis. This rotation then facilitates a rotor and stator generator system.

winter blend gasoline: A gasoline variant higher in volatile hydrocarbons, particularly butane, a compound that is both abundant and octane enhancing, to reduce gasoline's cost to consumers.

winter oxyfuels program: Properly called the State Winter Oxygenated Fuel Program, this refers to the EPA's implementation of the Clean Air Act, requiring oxygenated gasoline in areas where carbon monoxide levels during the winter exceed EPA standards. The type of oxygenate presently used in the U.S. is ethanol.

winterized diesel: An alternate diesel fuel blend designed to improve cold weather performance. This is achieved either by blending lower density distillates (kerosene) or additives to reduce pour point.

wireline: Unlike heavy-duty drilling pipe, a single cable can be used to lower a variety of downhole instruments, such as a perforating gun or well logging sensors.

working gas: The volume of natural gas stored in underground reservoirs in excess of its cushion gas. Working gas is continuously removed from and added back into the reservoir according to seasonal demand patterns.

workover rig: Equipment, typically mounted on a truck and smaller than a drilling rig, that carries out the tasks associated with the recompletion or rework of a well.

Z

zonal isolation: Refers to the use of cemented surface casing through certain subsurface formations (or zones), commonly used to protect water aquifers and noncommercial hydrocarbon-bearing reservoirs.

Made in United States
Orlando, FL
04 November 2026

90207780R00055